New Frontiers in Translation Studies

Series Editor

Defeng Li, Center for Studies of Translation, Interpreting and Cognition,
University of Macau, Macao SAR, China

Translation Studies as a discipline has witnessed the fastest growth in the last 40 years. With translation becoming increasingly more important in today's glocalized world, some have even observed a general translational turn in humanities in recent years. The New Frontiers in Translation Studies aims to capture the newest developments in translation studies, with a focus on:

- Translation Studies research methodology, an area of growing interest amongst translation students and teachers;
- Data-based empirical translation studies, a strong point of growth for the discipline because of the scientific nature of the quantitative and/or qualitative methods adopted in the investigations; and
- Asian translation thoughts and theories, to complement the current Eurocentric translation studies.

Submission and Peer Review:

The editor welcomes book proposals from experienced scholars as well as young aspiring researchers. Please send a short description of 500 words to the editor Prof. Defeng Li at Springernfits@gmail.com and Springer Senior Publishing Editor Rebecca Zhu: Rebecca.zhu@springernature.com. All proposals will undergo peer review to permit an initial evaluation. If accepted, the final manuscript will be peer reviewed internally by the series editor as well as externally (single blind) by Springer ahead of acceptance and publication.

Jun Pan · Sara Laviosa
Editors

Corpora and Translation Education

Advances and Challenges

Springer

Editors
Jun Pan
Department of Translation, Interpreting and Intercultural Studies
Hong Kong Baptist University
Kowloon, Hong Kong, China

Sara Laviosa
Department of Humanistic Research and Innovation
University of Bari Aldo Moro
Moro, Bari, Italy

ISSN 2197-8689 ISSN 2197-8697 (electronic)
New Frontiers in Translation Studies
ISBN 978-981-99-6588-5 ISBN 978-981-99-6589-2 (eBook)
https://doi.org/10.1007/978-981-99-6589-2

This Springer imprint is published by the registered company Springer Nature Singapore Pte Ltd.
The registered company address is: 152 Beach Road, #21-01/04 Gateway East, Singapore 189721, Singapore

Acknowledgements

We appreciate the support of the Language Fund under Research and Development Projects 2018–19 of the Standing Committee on Language Education and Research (SCOLAR), Hong Kong SAR, for making this publication possible. We would also like to thank the anonymous reviewers of the volume for their detailed and constructive feedback that helped much in improving the quality of the volume.

Contents

Learner Corpora

Editors and Contributors

About the Editors

Jun Pan is an associate professor in the Department of Translation, Interpreting, and Intercultural Studies at Hong Kong Baptist University, where she also holds the positions of the associate dean (Research) of the Faculty of Arts and the associate head of the Department. She serves as the co-editor of *Bandung: Journal of the Global South* and the review editor of *The Interpreter and Translator Trainer*. Her research interests lie in learner factors in interpreter training, corpus-based interpreting/translation studies, digital humanities and interpreting/translation, interpreting/translation and political discourse, professionalism in interpreting, etc. Her recent work includes a 6.5-million-word corpus on Chinese/English political interpreting and translation (https://digital.lib.hkbu.edu.hk/cepic/). She is also the president of the Hong Kong Translation Society.

Sara Laviosa is an associate professor in English and Translation at the University of Bari 'Aldo Moro', Italy. She has published extensively in international journals and collected volumes and is the author of *Corpus-Based Translation Studies* (Rodopi/Brill, 2002) and *Translation and Language Education* (Routledge, 2014). She is the co-author (with A. Pagano, H. Kempannen and M. Ji) of *Textual and Contextual Analysis in Empirical Translation Studies* (Springer, 2017). Her recent publications include *The Routledge Handbook of Translation and Education* (co-edited with M. González Davies, 2020), *The Oxford Handbook of Translation and Social Practices* (co-edited with M. Ji, 2020), *CTS Spring-Cleaning: A Critical Reflection*, Special Issue of *MonTI* (co-edited with M. Calzada Peréz, 2021), and *Recent Trends in Corpus-based Translation Studies*, Special Issue of *Translation Quarterly* (co-edited with Kanglong Liu, 2021).

Contributors

Michael Carl Kent State University, Kent, USA

Mark Davies English-Corpora.org, Springville, USA

Marie Escribe Universitat Politècnica de València, Valencia, Spain

Gaetano Falco Università degli Studi di Bari Aldo Moro, Bari, Italy

Devin Gilbert Utah Valley University, Orem, USA

Sara Laviosa Università degli Studi di Bari Aldo Moro, Bari, Italy

Dechao Li The Hong Kong Polytechnic University, Hong Kong, China

Kanglong Liu The Hong Kong Polytechnic University, Hong Kong, China

Ruslan Mitkov Lancaster University, Lancaster, England

Adam Obrusnik Masaryk University, Brno, Czechia

Jun Pan Hong Kong Baptist University, Kowloon Tong, Hong Kong, China

Mark Shuttleworth Hong Kong Baptist University, Kowloon, Hong Kong, China

Yanfang Su The Hong Kong Polytechnic University, Hong Kong, China

Honghua Wang The Hang Seng University of Hong Kong, Siu Lek Yuen, Hong Kong, China

Billy Tak Ming Wong Hong Kong Metropolitan University, Ho Man Tin, Hong Kong, China

Longhui Zou Kent State University, Kent, USA

Introduction

Jun Pan and Sara Laviosa

1 Premises

Corpora are large collections of language data that provide authentic examples of language use. They enable translators and translation students to tap into existing translations and develop a better understanding of the nuances of the source language and/or the target language. The use of corpora in translation education has, in recent years, become widespread thanks to the increasing availability of digital resources and the advances made in technical platforms, which make such resources more and more easily accessible.

Consequently, there has been a growing interest in exploring the use of corpora in translation education. Corpora can provide students with valuable linguistic resources and learning tools by means of which they can gain a deeper understanding of cross-linguistic differences and similarities that have a bearing on the translation process. As the field continues to evolve, it is highly likely that corpora will play an increasingly important role in the education and training of future translators.

The existing literature on corpora and translation education has explored various aspects of corpus use in translation learning and teaching. Some studies have examined the general landscape delineated by corpus-aided translation education (e.g., Laviosa and González-Davies 2020). Other studies have explored the theoretical underpinnings of corpus-based studies of translation as a subfield of the discipline of translation studies as a whole (e.g., Laviosa 2002; Hu 2016; Hu and Kim 2020).

There are also introductory books to corpus resources. For instance, Zanettin's (2014) research work provides an overview of different types of corpora and their

---

J. Pan (✉)
Hong Kong Baptist University, Kowloon Tong, Hong Kong, China
e-mail: janicepan@hkbu.edu.hk

S. Laviosa
Università degli Studi di Bari Aldo Moro, Bari, Italy
e-mail: sara.laviosa@uniba.it

J. Pan and S. Laviosa (eds.), *Corpora and Translation Education*, New Frontiers in Translation Studies, https://doi.org/10.1007/978-981-99-6589-2_1

potential applications in translation studies. Moreover, some studies have offered guidance on how to integrate corpora into translation education (Beeby et al. 2009; Zanettin et al. 2003) or translation practice (Mahadi et al. 2010). More recently, there have been case studies of the ways corpora are applied in the teaching of translation and their potential pitfalls (e.g., Liu 2020).

Overall, the existing literature on corpora and translation education reflects the broad range of applications and potential benefits deriving from the use of corpora for pedagogic purposes. Nevertheless, we have not revealed the full picture of the diverse and multifarious field of corpus-based or corpus-driven translation education, wherein corpora are used as a means of enhancing the teaching and learning of translation. There is also a need to introduce recent resources, obtain new empirical findings, and develop innovative perspectives on machine learning.

Against this background, this edited volume has been contemplated as a fruitful outcome of the International Symposium on Corpora and Translation Education held online in June 2021. The symposium aimed to address and bring together views on the advances and challenges related to corpora and translation education, including recent developments in the creation of resources for translation learner corpus research, the applications of corpora in translation education, the challenges posed by the collection and exploration of corpus-based learner data, the technical aspects of corpus building, good examples of corpus development, and online data platform design. After a two-day discussion, the participants in the forum wrote their papers about the role played by corpora in translation education today and how they can be used to enhance the teaching and learning of translation.

The proposed volume brings all of the aforementioned issues together, and provides an updated and comprehensive picture of both the theoretical underpinnings and practical resources of the robust field of corpus-based translation education, with the integration of perspectives on machine learning and post-editing.

2 This Volume

The edited volume covers a range of topics related to the use of corpora in translation education, including their position in corpus-based translation studies, the relationship of corpora, machine learning and post-editing, recent advances in learner corpora development, and the integration of corpora into translation pedagogy. The book also discusses the challenges and limitations of using corpora in translation education and proposes potential solutions. The volume can be neatly divided into four parts:

- Part I. Overview
- Part II. Corpora, machine learning, and post-editing
- Part III. Corpora and translation teaching
- Part IV. Learner corpora

In the opening chapter, also Part I of the volume, **Sara Laviosa** and **Gaetano Falco** present an overview of the past, present, and future of corpora and translation education. They start with an outline of the origin of Corpus Linguistics and Corpus-based Translation Studies, and then move on to survey corpus-based translator education from the late 1990s till the present day. Their definition of corpus-based translator education, i.e., "the use of bilingual comparable corpora or monolingual target language corpora as sources of data for experimental or classroom-based observational studies" (in this volume) helps to pinpoint the theoretical scope of the volume. In conclusion, the authors, based on their reflections on recent practices, point to future directions in the development of this subfield of study.

Part II of the volume consists of two chapters that focus on the innovative perspective of machine learning and post-editing. In the chapter "Applying Incremental Learning to Post-editing Systems: Towards Online Adaptation for Automatic Post-editing Models", **Marie Escribe** and **Ruslan Mitkov** address the difficulties encountered in developing tools and models to fully automate the post-editing process. The authors trained automatic post-editing models in a traditional setting and updated in both batch and online modes, with a view to analysing the performance of incremental adaptations in different systems, domains and language pairs. They developed an interactive functionality allowing for dynamic post-editing. The results of the study confirm the difficulties faced by the task of automatic post-editing. The authors make practical recommendations such as experimenting with more data (possibly synthetic corpora) and different environmental variables.

In the chapter "Integrating Trados-Qualitivity Data to the CRITT TPR-DB: Measuring Post-editing Process Data in an Ecologically Valid Setting", **Longhui Zou**, **Michael Carl** and **Devin Gilbert** approach post-editing process data by integrating Trados-Qualitivity Data into the CRITT Translation Process Research Database (CRITT TPR-DB), with the purpose to increase the ecological validity of TPR. This tool enables scholars, student translators and translation instructors to track different subjects' translation behaviour, increase their awareness of productivity, and characterise their translation styles. The authors conclude the chapter pointing out the great potentials of the data collected and processed via the new Qualitivity–TPR-DB interface.

Part III follows the prominent vein of corpus-based translation education, and features three chapters detailing the application of corpora in translation teaching. In the chapter "Creating and Using "Virtual Corpora" to Extract and Analyse Domain-Specific Vocabulary at English-Corpora.org", **Mark Davies**, tapping into his mega-sized English-Corpora.org, offers a step-by-step guide to the creation and use of "virtual corpora" for translators and teachers of English for Specific Purposes to extract and analyse domain-specific lists of words and multiword expressions. Additionally, Davies introduces means to limit searches of a larger corpus (like COCA or NOW or iWeb) to a particular virtual corpus and even compare results from different virtual corpora side-by-side.

In the chapter "Working with Corpora in Translation Technology Teaching: Enhancing Aspects of Course Design", **Mark Shuttleworth**, from a novice perspective of computer-assisted translation (CAT), compares and discusses the application of tools for terminology extraction, including Memsource and Sketch Engine. He then continues to introduce sources, where large-size online corpora could be downloaded for Translation Memory (TM) build-up, such as the OPUS website.

In the final chapter of Part III, **Kanglong Liu**, **Yanfang Su**, and **Dechao Li** present an empirical study of the application of a large-scale parallel corpus in translation teaching. Based on both qualitative and quantitative data, the study shows that using a parallel corpus in general brings benefits to students, raising their awareness of translation problems and strategies, as well as enhancing resourcefulness. The use of a parallel corpus was also generally well received by students. At the end of the chapter, the authors also discuss the challenges posed by corpus-assisted translation teaching, such as corpus design and pedagogical planning.

Part IV of the volume highlights another aspect of corpus-based translation education, which is learner corpora. In the chapter "Data Acquisition and Other Technical Challenges in Learner Corpora and Translation Learner Corpora", **Adam Obrusnik**, the developer of Hypal and Hypal4MUST software, discusses the technical aspects of data acquisition for learner and translation corpora. Using Hypal and Hypal4MUST software tools as examples, the author illustrates how the burden of assembling metadata can be divided among 3 key personas—the researcher, the teacher, and the student.

In the concluding chapter, **Jun Pan**, **Billy Tak Ming Wong**, and **Honghua Wang** introduce their collective efforts in developing and analysing a large-size translation learner corpus, i.e., the Hong Kong subset of the Multilingual Student Translation (MUST) corpus, which has been designed and compiled as part of an international research initiative. The study illustrates the challenges posed by thickly annotating learner data, its potential, and how the findings based on this data can bring about positive changes in curriculum design and educational policies.

The volume adopts a practical perspective, which is supported by ample evidence collected from case studies that provide real-life examples to help readers apply the concepts and methods discussed in each chapter. Readers will gain a deeper understanding of the potential benefits and challenges of using corpora in translation education, and receive practical guidance on how to incorporate corpora into their teaching practice.

Aiming to explore the role of corpora in translation education and how they can be used to enhance the teaching and learning of translation, the volume covers a range of topics related to the use of corpora in translation education, including corpus-based translation studies, the integration of corpora into translation pedagogy, and the assessment of corpus-based translation skills. It also addresses the challenges and limitations of using corpora in translation education and proposes potential solutions. These insights also hold the potential to guide the development of more advanced Artificial Intelligence systems for languages. This, in turn, can shape the future of translation education, making resources more easily accessible and fostering a more data-driven approach in the field.

In general, the volume will have an impact on a number of related disciplines:

(1) Translation Studies: The book will address, first and foremost the concerns of the discipline of Translation Studies. It looks into how corpus linguistics can be used to enhance translation education. It also provides examples of different uses of corpora in translation practice.
(2) Linguistics: The book explores the theoretical underpinnings of corpus linguistics, a subfield of linguistics, and its applications in translation education. The theoretical and methodological issues discussed can shed light on linguistics at large.
(3) Education: The book illustrates pedagogical resources and tools for teaching and learning translation, and discusses the impact of technology on translation education. The discussion can feed back into the general discussion of education, especially in the age of artificial intelligence wherein technology plays an increasingly important role.
(4) Computer Science: The book's discussion of the technical aspects of building and using corpora, and the role of machine translation in translation education, makes it also relevant to computer science.

References

Beeby, Allison, Patricia Rodríguez Inés, and Pilar Sánchez-Gijón, eds. 2009. *Corpus Use and Translating*. Amsterdam: John Benjamins.

Hu, Kaibao. 2016. *Introducing Corpus-Based Translation Studies*. Series Editor: Defeng Li. Singapore: Springer.

Hu, Kaibao, and Kim Kyung Hye, eds. 2020. *Corpus-Based Translation and Interpreting Studies in Chinese Contexts*. Singapore: Springer.

Laviosa, Sara. 2002. *Corpus-Based Translation Studies: Theory, Findings, Applications*. Amsterdam: Rodopi/Leiden: Brill.

Laviosa, Sara, and Maria González-Davies. 2020. *The Routledge Handbook of Translation and Education*. London: Routledge.

Liu, Kanglong. 2020. *Corpus-Assisted Translation Teaching: Issues and Challenges*. Series Editor: Defeng Li. Singapore: Springer.

Mahadi, Tengku Sepoa, Tengku Mahadi, Helia Vaezian, and Mahmoud Akbari. 2010. *Corpora in Translation: A Practical Guide*, vol. 120. Bern: Peter Lang.

Zanettin, Federico. 2014. *Translation-Driven Corpora: Corpus Resources for Descriptive and Applied Translation Studies*. London: Routledge.

Zanettin, Federico, Silvia Bernardini, and Dminic Stewart, eds. 2003. *Corpora in Translator Education*. Manchester: St. Jerome Publishing.

Jun Pan is Associate Professor in the Department of Translation, Interpreting, and Intercultural Studies at Hong Kong Baptist University, where she also holds the positions of Associate Dean (Research) of the Faculty of Arts and Associate Head of the Department. She serves as Co-editor of *Bandung: Journal of the Global South* and Review Editor of *The Interpreter and Translator Trainer*. Her research interests lie in learner factors in interpreter training, corpus-based interpreting/translation studies, digital humanities and interpreting/translation, interpreting/translation and political discourse, professionalism in interpreting, etc. Her recent work includes a 6.5 million-word corpus on Chinese/English political interpreting and translation (https://digital.lib.hkbu.edu.hk/cepic/). Dr. Pan is also President of the Hong Kong Translation Society.

Overview

Corpora and Translator Education: Past, Present, and Future

Sara Laviosa and Gaetano Falco

1 Introduction

In this section, we provide the definition of key terms and concepts in order to delineate our specific field of enquiry. Firstly, corpus-based translator education denotes an area of research within Corpus-based Translation Studies (CTS), which adopts and develops the methods and tools of Corpus Linguistics (CL) to analyse translation practices for applied purposes, most notably teaching methods, testing techniques, and curriculum planning. Corpus-based Translation Studies (CTS) is an area of research that adopts and develops the methodologies of Corpus Linguistics (CL) to analyse translation practices for theoretical, descriptive, and applied purposes. Corpus Linguistics (CL) is an approach to the empirical study of language that relies on the use of corpora. The key words are therefore approach, empirical, and corpora. In turn, corpora are collections of authentic unabridged texts or whole sections of text held in electronic form and assembled according to specific design criteria "to represent, as far as possible, a language or language variety as a source of data for linguistic research" (Sinclair 2005: 16). The key aspects to bear in mind in this definition are authentic, electronic form, design criteria, and representativeness.

What is distinctive about the empirical approach to the study of natural languages adopted by Corpus Linguistics is the integration of four elements: linguistic theory, observable linguistic data, research methods and tools, and description of language use. The interrelationship of these elements is evident and relevant when we undertake corpus-based research. CL studies involve a continual process that starts with the formulation of testable hypotheses about language use, which are based on linguistic theory. These hypotheses are then investigated through quantitative and

S. Laviosa (✉) · G. Falco
Università degli Studi di Bari Aldo Moro, Bari, Italy
e-mail: sara.laviosa@uniba.it

G. Falco
e-mail: gaetano.falco@uniba.it

J. Pan and S. Laviosa (eds.), *Corpora and Translation Education*, New Frontiers in Translation Studies, https://doi.org/10.1007/978-981-99-6589-2_2

qualitative analyses of corpus data. The analysis in turn provides empirical findings that are systematically organised in new descriptions of language use, which then feed into linguistic theory where descriptions are accounted for. On the basis of these explanations, new hypotheses are put forward and the whole cycle starts all over again.

What is distinctive about research into corpus-based translator education—which finds its place within Applied CTS—is the use of bilingual comparable corpora or monolingual target language corpora as sources of data for experimental or classroom-based observational studies. The main goal of this variegated body of research is to enhance the acquisition of translation skills and target language competence, the improvement of translation quality assessment, and the development of translation aids. Application-driven research such as this is nurtured mainly by students' learning needs, which are assessed and addressed within the training environment. Hypotheses are put forward and tested in the classroom by scholar-teachers who are familiar with the achievements in theoretical and descriptive translation studies. In turn, the findings obtained from applied research are accounted for in theoretical terms and feed into description, theory, and teaching methodologies. In so doing, the gap between theory and practice is narrowed in the actual day-to-day training of professional translators.

2 Corpus Linguistics

As we know it today, CL began to take shape in the 1980s, when the term "corpus linguistics" was coined by Jan Aarts and first appeared in the title of a collected volume, *Corpus Linguistics: Recent Developments in the Use of Computer Corpora in English Language Research* (Aarts and Meijs 1984). During the 1980s multi-million-word corpora of written and spoken English were compiled. Over the same decade, there was a development from post-Bloomfieldian American Structuralism underpinning the resource-driven goals of early corpus linguists in the 1960s to the neo-Firthian theoretical framework. According to the post-Bloomfieldian approach of the 1940s and 1950s—which drew on the research work of Leonard Bloomfield (1887–1949)—language consists of phonological, morphological, and syntactic structures. The study of language was characterised by strict empiricism, and the grammar of a language was unveiled by the direct observation of a corpus of data through "discovery procedures". John Rupert Firth (1890–1960) advocated the study of authentic language use in texts. "[T]he complete meaning of a word", he maintained, "is always contextual, and no study of meaning apart from a complete context can be taken seriously" (Firth 1957: 37). Firth proposed to explore the meaning of words through their distribution in different contexts and their habitual collocations, which "are quite simply the mere word accompaniment" (1957: 11). Firth's view of language is functional, since any utterance is regarded as "a way of acting on other people and influencing one's environment" (1957: 36). From the 1960s onwards,

Firth's contextual theory of meaning was developed by the British linguists Michael Halliday and John McH. Sinclair.

In the 1980s, large corpora began to be used for theory-driven research and for creating a new generation of usage-based dictionaries aimed at advanced learners of English. First and foremost, the Collins-Birmingham University International Language Database (COBUILD) was created in the 1980s under the direction of John McH. Sinclair and was designed specifically for linguistic research and lexicography. In order to build this innovative language database new technology, i.e. the optical character reader was used to read large quantities of printed text and access much material already available in machine-readable form. Also, during the 1980s, the Longman/Lancaster English Corpus was developed by Della Summers (at the publishing house of Longman) and Geoffrey Leech (Lancaster University). This is a lexicographic corpus that used material published since 1900 in both British and American English. It comprises 30 million words of written text. Furthermore, the International Corpus of English (ICE), coordinated by Sidney Greenbaum at University College London, began in the late 1980s with the aim of representing spoken, printed, and manuscript samples of English in countries where it is a first or official second language, each national component being a million running words.

In the following decade, the British National Corpus (BNC) was compiled between 1991 and 1994 with the support of Longman, Oxford University Press, Chambers Harrap, the Oxford University Computing Service, the University of Lancaster, the British Library, and the Department of Trade and Industry. The BNC consists of 100 million words of British English—90 million of written text and 10 million of spoken text—sampled from 1960 onwards. Meanwhile, after COBUILD, the Bank of English project was set up in 1991 by Collins and the University of Birmingham. It was the first open-ended monitor corpus and now contains 650 million words. The large-scale study of written and spoken language usage made possible by the advent of large computerised corpora gave rise to "a new view of language" (Sinclair 1991: 1) and "a new way of thinking about language" (Leech 1992: 106). The main tenets of CL can be summarised as follows:

- Language is viewed as a social phenomenon which reflects, constructs, and reproduces culture.
- Language in use is systematically heterogeneous; texts are therefore studied comparatively across corpora that are intended to represent different language varieties.
- Language in use involves both routine and creative processes; typicality plays a socializing role.
- The aim of studying language in corpora is to describe and explain the observed phenomena, not to predict what some other corpus may contain.
- Linguistics is essentially a social science and an applied science.

Against this background, Mona Baker (1993) proposed to use corpora for the empirical study of translation in a paper published in a collected volume in honour of John McH. Sinclair (Baker et al. 1993). In that contribution, Baker (1993: 243) claimed that "[t]he availability of large corpora of both original and translated text, together with the development of a corpus-driven methodology [would] enable translation scholars to uncover the nature of translated text as a mediated communicative event". Differently from two previous corpus studies of English-Swedish translation, which were intended to improve translation practice (Gellerstam 1986; Lindquist 1989), what Baker proposed in the early 1990s was a research programme conceived within the target-oriented perspective of Descriptive Translation Studies (DTS).

3 Corpus-Based Translation Studies

The first collection of papers devoted to Corpus-based Translation Studies was published in a special issue of *Meta* entitled *L'Approchee Basée sur le Corpus/ The Corpus-Based Approach* and guest edited by Sara Laviosa (Laviosa 1998). The theoretical papers outlined the scope, object of study, and methodology of the emergent corpus-based approach. The empirical papers reported on the results of contrastive studies and the investigation of the product and process of translation. The pedagogical papers reported on the use of corpora for translator training. It was in Miriam Shlesinger's paper (1998) that the terms "corpus-based translation studies" and "corpus-based interpreting studies" first appeared in the literature after being coined by Sara Laviosa-Braithwaite (1996) in her doctoral thesis. In her contribution to the special issue, "Corpus-based Interpreting Studies as an Offshoot of Corpus-based Translation Studies", Shlesinger (1998) looked at the specific problems and benefits arising from the application of a corpus-based methodology to investigate interpreted texts as distinct instantiations of oral discourse and to identify the regular patterns of language use that distinguish interpreting from written translation. The concluding paper by Maria Tymoczko, "Computerised Corpora and the Future of Translation Studies", drew on the insights provided by the theoretical, descriptive, and pedagogical papers and pointed out that:

> One of the most encouraging aspects of the pioneering studies of CTS is the way that seemingly technical and theoretical interrogations come to have practical potential and immediate applicability, not only for the teaching of translation but for the work of the practising translator as well. (1998: 658)

The papers presented in the special issue of *Meta* illustrate some of the main lines of enquiry that would be developed in the years that followed. These can be grouped under three main headings: theory, description, applications. Within Applied CTS, an area of study that is particularly relevant to our present discussion, it is worth mentioning the early research work of Federico Zanettin (1998) and Lynne Bowker (1998). Zanettin demonstrated how small bilingual comparable corpora, designed and assembled in his undergraduate translation classroom at the University of Bologna, can be useful to compare words and phrases that have strong resemblance across languages such as proper names, cognates or lexicographic translation equivalents. The comparisons that were carried out in class showed, e.g. how Italian and English broadsheet newspapers had different ways of naming leading political figures. So, *Francois Mitterand* and *Mitterand* were used in Italian, while English preferred to use the appropriate title, i.e. *President Mitterand*, *President Francois Mitterand* or *Mr Mitterand*. Also, equivalent verbs typically used to introduce direct and reported speech were found to have different frequencies of occurrence as well as different lexico-grammatical profiles in the two languages. Even cognates such as *prezzi* and *prices* were discovered to have dissimilar collocational and colligational patterns. Of course, today we take these contrastive features for granted, but they were a novelty at the time. The classroom-based investigations illustrated by Zanettin, which were framed within a student-centred, discovery learning approach, were and, arguably still are, valuable in yielding data that enrich contrastive analysis and enhance translation skills.

Still within a pedagogic perspective, Bowker addressed two problems usually encountered by translator trainees in specialised subject domains. These are: terminological errors resulting from poor subject-specific knowledge and lack of specialised writing skills in language A, the language they translate into. In her paper, Bowker reported on an experiment she had carried out with her fourth-year undergraduate L1 English students at Dublin City University. The students produced two translations from French into English of two semi-specialised passages on optical scanners. One translation was completed with the aid of bilingual and monolingual dictionaries together with other reference materials (e.g. manuals and brochures). The other translation was carried out with a bilingual dictionary and a 1.4 million-word specialised monolingual corpus of English articles on optical scanners, which was compiled from Computer Select on CD-ROM and analysed with the 1st version of Mike Scott's WordSmith Tools. The findings were quite revealing. Corpus-aided translations were of higher quality with regard to subject-field understanding, correct term choice, and correct phraseology. There was no difference as to grammar and register. Summing up, in the late 1990s CTS was given a name. It covered three main research domains, i.e. theory, description, applications. Alongside CTS, the seeds were planted for the development of Corpus-based Interpreting Studies (CIS), which flourished from then on (see Russo et al. 2018).

The first decade of the new millennium is marked by two international conferences entirely devoted to CTS. The first was held in Pretoria from 22 to 25 July 2003 and the theme was "Corpus-based Translation Studies: Research and Applications". As Alet Kruger (2004: 2) recalls in the Editorial of the special issue of *Language Matters: Studies in the Languages of Africa*, where a selection of the papers presented at the conference were published,

> this was the first conference of its kind in South Africa and the first focusing only on CTS in the world. The aim of the conference was to consider ways in which corpora could be used to develop novel and challenging perspectives in the discipline, as well as ways in which they could support research outside the mainstream hegemonic research cultures.

The second conference was hosted in Shanghai from 31 March to 3 April 2007 and the theme was "Corpora and Translation Studies". Also, in the early 2000s, two monographs on CTS were published in England (Laviosa 2002; Olohan 2004) together with the first collection of papers on translator education (Zanettin et al.

2003). Thirteen years on from the publication of the special issue of *Meta* and seven years on from the publication of the special issue of *Language Matters*, another collection of papers on CTS was published in 2011. As the editors point out in the Introduction,

> The articles in this volume are written by many of the leading international figures in the field. They provide an overall view of developments in corpus-based translation (and interpreting) studies and also specific case studies of how the methodology is employed in specific scenarios, such as contrastive studies, terminology research, and stylistics. (Kruger et al. 2011: 1)

The lines of enquiry represented in this collection of essays are: theory, description, applications, and tools. In this volume Bowker (2011) focuses on translation aids and curriculum design. She surveys the development of Computer-Assisted Translation (CAT) tools such as corpora and aligned pairs of texts stored in specially designed databases that are processed by translation memory systems and automatic term extractors. More specifically, Bowker looks at:

- How technology has been integrated in the professional practice of terminologists.
- How translators use the computer-based resources created by terminologists such as term-banks.
- How corpora have influenced the process and product of terminological research carried out by terminologists and translators.

Then, on the basis of this survey, Bowker discusses the implications for terminology training, which, in her view, should be a core component of translator training programmes together with training in technology. While Bowker addresses issues relevant to translator education, Zanettin (2011) describes the multiple layers of annotation based on XML/TEI standards for bidirectional parallel corpora. And Luz (2011) describes the web-based infrastructure for creating and sharing dynamic and widely accessible corpora such as the Translational English Corpus (TEC) developed at the Centre for Translation and Intercultural Studies (CTIS), University of Manchester.

4 Corpus-Based Translator Education

And now we come to the present day. As sketched out in Fig. 1, Applied CTS has made inroads into all three research areas of Applied Translation Studies, namely translator training, translation aids, and translation quality assessment. In turn, corpus-based translator training includes teaching methods, testing techniques, and curriculum design. It is worth pointing out that Applied CTS engages in a constructive dialogue with Descriptive CTS, where systematic empirical investigations into the product, process, and function of translation, together with the development of appropriate research methodologies, have a bearing on all the ambits of scholarly enquiry and practice subsumed under Applied Translation Studies (or Applied Extensions of Translation Studies).

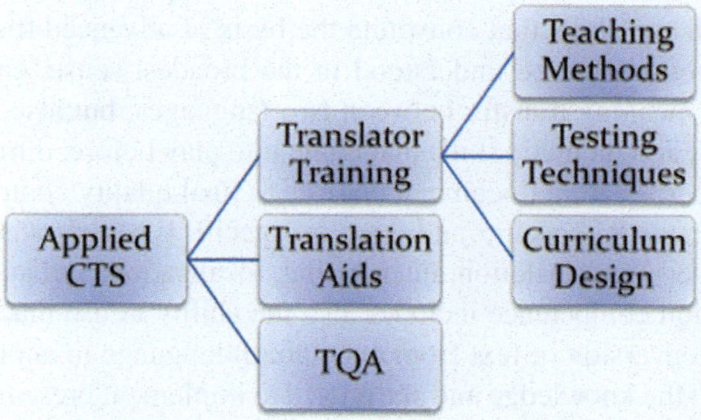

Fig. 1 Applied corpus-based translation studies

In order to survey the state of the art of corpus-based teaching methods in translator training, in particular, it is important to introduce the *European Master's in Translation Competence Framework 2017* (Toudic and Krause 2017). The EMT is a network of Master's level study programmes that was developed in 2009 by higher education institutions in partnership with the European Commission's Directorate General for Translation (DGT). The *EMT Competence Framework 2017* was drawn out in response to three main developments that have occurred in the provision of translation services in the last decade. These developments are: (a) the impact of technology, (b) the continuing expansion of English as lingua franca, and (c) the role of Artificial Intelligence and social media in communication. The framework views translating as a process to meet individual, societal, or institutional needs.

The aim of the *EMT Competence Framework 2017* is to consolidate and enhance employability of graduates with Master's degrees in translation throughout Europe. It considers translation a multi-faceted profession and recommends that translator training at Master's degree level should equip students not only with a deep understanding of all the processes taking place when conveying meaning from one language to another, but also with the ability to perform and provide a translation service in line with the highest professional and ethical standards. The framework defines five complementary areas of competence, all equally important: (a) language and culture (transcultural and sociolinguistic awareness and communicative skills); (b) translation (strategic, methodological and thematic competence); (c) technology (tools and applications); (d) personal and interpersonal; (e) service provision.

We will now expound each competence area in turn. Language and culture includes all the general and language-specific linguistic, sociolinguistic, cultural, and transcultural knowledge and skills that constitute the basis of advanced translation competence. Translation should be understood in the broadest sense, encompassing not only the actual meaning transfer between two languages, but also all the strategic, methodological, and thematic skills that come into play before, during, and after the transfer phase per se, from document analysis to final quality control procedures in domain-specific, media-specific, and situation-specific types of translation. The latter include public service translation, interpreting, localisation, and audio-visual translation. Translation competence includes also the ability to use machine translation, the automatic conversion of text from one natural language to another. Technology encompasses all the knowledge and skills used to implement present and future technologies during the different phases of the translation process. It also includes basic knowledge of machine translation and the ability to utilise it when needed. As we can see in the list reproduced in Table 1, the ability to use computerised corpora as translation aids is an integral part of this area of competence together with the ability to use search engines, text analysis tools, and CAT tools.

The personal and interpersonal area of competence comprises all the so called "soft skills", namely planning and managing time, stress and workload; complying with deadlines, instructions, and specifications; use of social media; self-evaluation; and collaborative learning. Finally, service provision covers all the skills relating to the provision of language services in a professional context, from client awareness and negotiation through to project management and quality assurance.

Table 1 Technological knowledge and skills

• Use the most relevant IT applications, including the full range of office software, and adapt rapidly to new tools and IT resources
• Make effective use of search engines, corpus-based tools, text analysis tools, and CAT tools
• Pre-process, process, and manage files and other media/sources as part of the translation, e.g. video and multimedia files, as well as handle web technologies
• Master the basics of MT and its impact on the translation process
• Assess the relevance of MT systems in a translation workflow and implement the appropriate MT system where relevant
• Apply other tools in support of language and translation technology, such as workflow management software

Toudic and Krause (2017: 9)

5 Corpus-Based Translation Training: An Overview of Recent Studies

In this section, we provide a sample of recent studies concerning the integration of corpora in translator training as a result of the spread of technological tools and expertise in the translation profession. Specifically, we discuss two issues. First of all, we outline some of the recent uses of corpora in translation training. Secondly, we highlight some of the problems and drawbacks arising from corpus-based translation training, and suggest what, as far as we are concerned, could be potential solutions. Notably, major problems are allegedly the result of practical reasons which, as we suggest, can be tackled by integrating cognitive tools.

5.1 *Introducing New Trends in Translation Training*

As has been noted in Sect. 4, the majority of recent translation teaching projects carried out at university level in Europe are grounded on the set of skills and competences identified by the EMT network in response to the emerging needs of European universities and translation industries, as well as the socio-cultural and technological changes that have occurred over the last decade (Torres-Simón and Pym 2019). In this context, the European Qualifications Framework (EQF) claims that the learning outcomes of a translation training programme shall include theoretical knowledge, practical and technical skills, and social competences where the ability to work with others will be crucial. To achieve such a goal, the EMT research group recommends a quality-driven translation pedagogy whose ultimate goal is turning apprentice translators into professionals. Among other things, the pedagogy shall also include the training of students in the use of corpora for translation purposes. A review of the research articles describing the findings of the projects implemented in the field of translation training at university level enable us to identify at least three basic issues which characterise this quality-driven pedagogy in translation training.

We begin with outlining the socio-constructivist approach. The use of corpora in translation training is, first of all, part of a pedagogical approach which can be defined "socio-constructivist", since it rests on the role of interpersonal, intersubjective interaction among students in the construction of their knowledge (Kiraly 2000; Gonzáles-Davies 2004). As a matter of fact, the technological turn that has taken place at different levels of training reveals a shift from a static way of teaching, largely based on a transmissionist, teacher-centred approach, to a more dynamic, proactive way of learning, which involves a learner-initiated approach, in that students "become responsible for their own learning and the learning of others. Teachers are no more the authority who determine what is studied and who assess the quality of their students' work" (Atan 2012: 2). Within a socio-constructivist perspective, the translation classroom is seen as a replication of the real world context, in which different authentic activities take place, including collaborative work, information exchange, exploratory attitudes, and inquiry-based learning (Sessoms 2008). From a sociological perspective, collaborative work is based on the division of labour, where each student performs specific tasks ranging from translation project management to pre-editing, compilation of corpora, data mining and glossary construction, translation proper, post-editing, and revision.

In this framework, translator trainees become active builders of their own knowledge, they monitor and are responsible for their education process, improving and upgrading their skills via collaborative learning. Indeed, trainees do not act in isolation, but are part of a community, in which each individual engages in a collaborative, productive process, interacting with different stakeholders. This pedagogic approach entails a change in power relations between teachers and students, as the latter are not merely consumers of the teaching process, but are empowered to become prosumers, i.e. decisive actors in designing and planning translation activities and syllabi.

Syllabus design is the second issue addressed within a quality-driven pedagogy in translation training. As a result of the technological turn, most syllabi for the training of prospective professional translators incorporate modules intended to train students to use technological tools, i.e. CAT tools, machine translation, translation memories, and collaborative translation platforms that they need in order to improve their outcomes in terms of quality and time pressure. Recent studies have shown the importance of integrating corpora as well as corpus concordancers and other software suites for corpus analysis in syllabi for translation training (Wang 2011). In addition to modules intended to develop technological competence (Sikora 2014) and enable students to become acquainted with techniques and software used in professional translation—e.g. database management systems, translation memory systems, proofreading, revisions and post-editing techniques, technologies used in the processes of document production and management —the vast majority of syllabi designed for programmes for translation training also include modules on learning how to query corpora and compile DIY corpora. The aim of such syllabi is to combine new technological competences with translation together with personal and interpersonal competences. From this perspective, competence in translation technology is not just a matter of automatic work, but it enhances critical thinking, creativity, and methodology. Disregarding this key aspect of translation technology would imply widening the gap between theory and practice, which has always been one of the conundrums of Translation Studies.

It follows that technology-related tasks must be contextualised rather than separated from realistic experience. It also means restoring the original meaning of technology to its Greek etymology, i.e. words or discourse (*Lógos*) about art, skill, craft, and especially the principles or methods employed in making something or attaining an objective (*Techné*). On these grounds, what is the role of corpora in translation training? This is the third issue of our discussion about quality-driven translation pedagogy, which we refer to as teaching styles. With regard to corpus-based approaches to translation training, drawing upon Beeby et al. (2009), Frérot (2016) distinguishes between two teaching styles: corpus use for learning to translate and learning corpus use to translate.

The corpus use for learning to translate style engages teachers in the design of corpus-based translation-related tasks so that students focus on a particular translation issue and analyse a given set of preselected data. The second type of teaching style, learning corpus use to translate, is student-initiated. In this case, students play a central role as they are involved in tasks critical to their own training, such as designing and compiling DIY corpora as well as identifying strategies and tools to search the corpora by themselves. The students' direct involvement in these activities is an added value to their training in translation: they learn how to use corpora efficiently and strategically as far as problems related to terminology, collocational patterns, genre, and discourse are concerned, and, consequently, solve real-life translation problems (Frérot 2016).

To illustrate the difference between the two styles, we briefly present two cases. An example of a resource representative of the corpus use for learning to translate style is the European Commission platform which allows users to display EU legal documents in one to three official languages of the European Union (Fig. 2) and retrieve terms, e.g. "balance sheet item", in the Source Language (SL) and their translations in two Target Languages (TL) (Fig. 3).

Platforms like the European Commission's are interesting corpus-based resources, especially when specialised translation is taught. They can be used in class to extract terms, retrieve lexico-semantic patterns, and genre conventions as well as their translation equivalents. Over the last two decades the number of corpus-based platforms has grown significantly. Also, there has been an increasing interest in designing web-derived corpora which are domain-specific and very large in size. Among these, an outstanding example of corpus colossal, in terms of size, language variety, and subject differentiation, is Sketch Engine (Fig. 4).

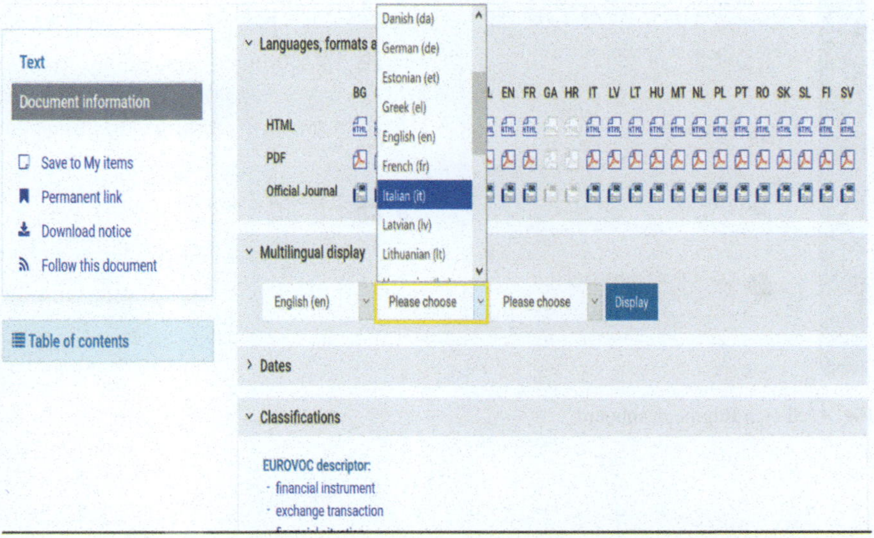

Fig. 2 European Commission Eur-Lex platform

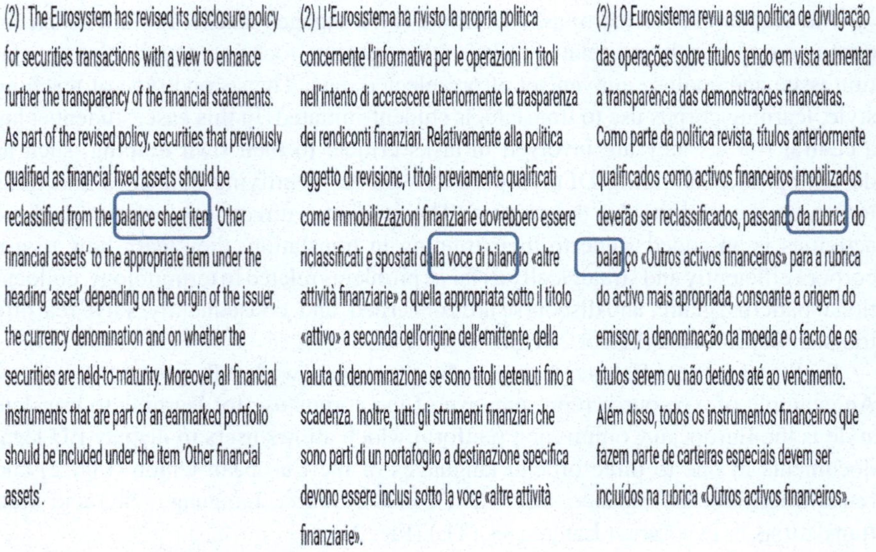

(2) | The Eurosystem has revised its disclosure policy for securities transactions with a view to enhance further the transparency of the financial statements. As part of the revised policy, securities that previously qualified as financial fixed assets should be reclassified from the balance sheet item 'Other financial assets' to the appropriate item under the heading 'asset' depending on the origin of the issuer, the currency denomination and on whether the securities are held-to-maturity. Moreover, all financial instruments that are part of an earmarked portfolio should be included under the item 'Other financial assets'.	(2) | L'Eurosistema ha rivisto la propria politica concernente l'informativa per le operazioni in titoli nell'intento di accrescere ulteriormente la trasparenza dei rendiconti finanziari. Relativamente alla politica oggetto di revisione, i titoli previamente qualificati come immobilizzazioni finanziarie dovrebbero essere riclassificati e spostati dalla voce di bilancio «altre attività finanziarie» a quella appropriata sotto il titolo «attivo» a seconda dell'origine dell'emittente, della valuta di denominazione se sono titoli detenuti fino a scadenza. Inoltre, tutti gli strumenti finanziari che sono parti di un portafoglio a destinazione specifica devono essere inclusi sotto la voce «altre attività finanziarie».	(2) | O Eurosistema reviu a sua política de divulgação das operações sobre títulos tendo em vista aumentar a transparência das demonstrações financeiras. Como parte da política revista, títulos anteriormente qualificados como activos financeiros imobilizados deverão ser reclassificados, passando da rubrica do balanço «Outros activos financeiros» para a rubrica do activo mais apropriada, consoante a origem do emissor, a denominação da moeda e o facto de os títulos serem ou não detidos até ao vencimento. Além disso, todos os instrumentos financeiros que fazem parte de carteiras especiais devem ser incluídos na rubrica «Outros activos financeiros».

Fig. 3 Parallel texts for "balance sheet item"

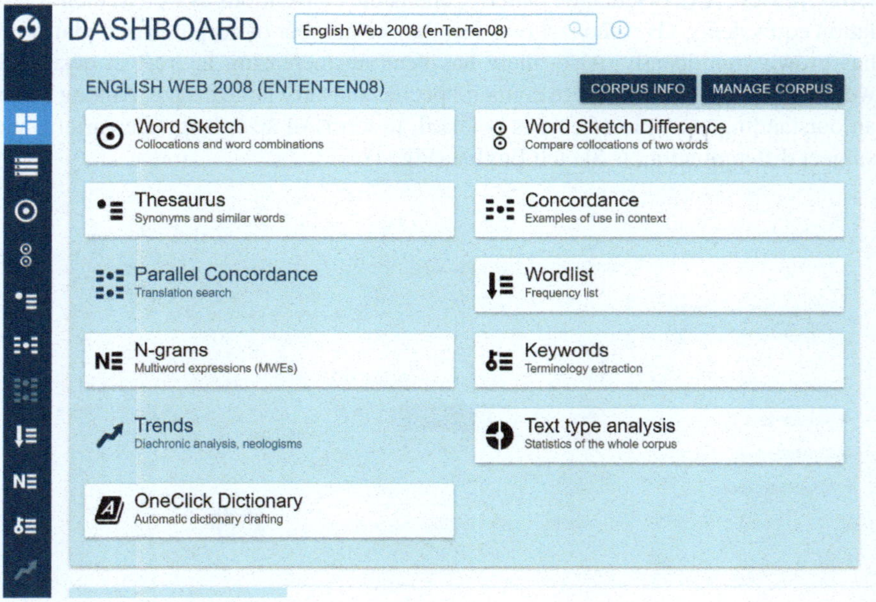

Fig. 4 Sketch Engine dashboard

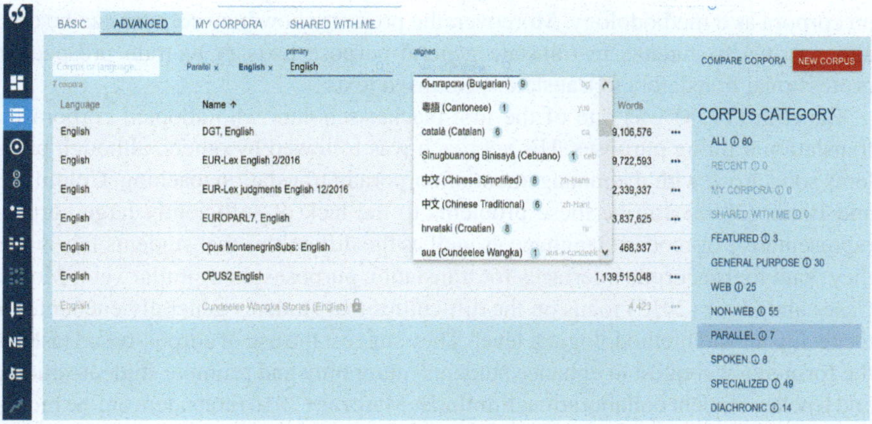

Fig. 5 Multilingual corpora in sketch engine

Sketch Engine collects a number of corpora in different natural languages, including corpora of parallel texts, such as the European Commission DGT, Eur-lex, Europarl, OPUS, the Bible, and the Quran corpora. Notably, it contains corpora representative of all the world languages, which can be organised by different criteria: mode, time, subject-domain. The corpora included in Sketch Engine are an important source of translation data and a valuable resource for creating bilingual glossaries of specialised terms (Fig. 5).

Both parallel EU documents and Sketch Engine can also be used as instruments for teaching students how to use corpora for translation purposes, which is the other learning style identified by Frérot (2016). To put it simply, in the case of the first style, the goal is training students to learn how to translate, i.e. translation is seen as a process, and corpora are just a resource to achieve such a goal. By contrast, in the case of the second style, the goal is training students to learn how to design and compile different types of corpora, including DIY parallel corpora, for language and translation purposes. In other words, translation is seen as a product and corpora are seen as a methodology, an approach with its theoretical rationale. Tellingly, these resources, including the Web, have been largely employed over the last years in many student-initiated translation teaching projects in which corpora are used for learning to translate and enhance student's creativity and autonomy. In the next section, we offer an overview of corpus-based translation training projects in higher education.

5.2 *Corpus-Informed Translation Training: An Overview*

The majority of translation training projects presented in this sub-section are representative of the two styles identified by Frérot (2016) since, on the whole, the projects are either focused on the use of corpora as instruments for learning to translate or on corpora as a methodology. Moreover, the projects show how corpora can be used either to teach students to translate general-purpose texts or to train prospective professional translators to translate specialised texts.

Zanettin (1998) was one of the first teacher-scholars who adopted corpora for translation training purposes. His approach was followed by others, although problems soon arose with the management of corpora in translation teaching. Colominas and Baida (2008) ascribe these problems to the lack of sufficiently large corpora representative of modern language as well as the difficulties that students face when they want to use corpus interfaces for translation purposes. In a similar vein, Molés-Cases and Oster (2015) focus on the difficulties that students generally encounter at the technical and methodological level. They suggest the use of corpus-based tasks in the form of a webquest to enhance students' autonomy and promote student–student and teacher-student collaboration. Similarly, Marczak (2016) considers online mono- and bilingual corpora an important methodological tool for data mining to build students' translation competence through telecollaboration.

Sikora (2014) considers corpora as one of the research and information mining competences that, along with the technological competences, such as CAT tools and translation memories, must be included in specialised translation syllabi since they improve translation competence in prospective professional translators. Relying on recent research on language learning and Translation Studies, Singer (2016), too, contends the importance of corpora for translation teaching. In particular, he proposes a data-driven learning approach for language learning within a translator training programme using a task-based learning (TBL) approach. For an exhaustive review of the most important studies and findings relevant to the important role that corpus linguistics plays in translation training, Neshkovska's article (2019) represents an outstanding resource.

Similar studies have been carried out in various language pairs and with corpora for special purposes. Alhassan et al. (2021) aim at improving language skills and translation competence using parallel Arabic-to English corpora, when they teach translation from Arabic into English to major students at an Omani private university. Gallego-Hernández (2015) presents a survey-based study that is relevant to the use of corpora by Spanish professional translators. Kübler (2011) discusses the important role played by corpus-based translation activities carried out by French-speaking students attending an MA course in specialised translation at the University Paris Diderot. She demonstrates that teaching students to translate texts in specialised domains—Earth Science in her specific case—requires students' awareness of what corpus use entails. In other words, looking for translation equivalents in parallel corpora is not enough; students should also be taught about corpora as a theory and methodology that can be exploited in the translation process.

Rodríguez's article (2016) offers an exhaustive overview of recent developments in the field of corpus use for translation training as a result of the technological turn. In particular, she shows how to exploit on-line bilingual concordancers and resources based on translation memories in order to translate scientific and technical texts from English into Spanish. Beside technology, her approach to translation training also relies on creativity, which she regards as a fundamental, cognitive activity underlying language learning and translation. Vigier-Moreno (2019) reports on the results of an initiative carried out at the Spanish University Pablo de Olavide of Seville (UPO), which was intended to train students to use monolingual corpora for translating specialised texts from English into Spanish, with a focus on phraseology and terminology. Lee et al. (2020) investigated the use of bilingual corpora as pedagogical tools for translation in the domain of technical writing. In particular, the corpus included Chinese translation of English patents. The use of a bilingual specialised corpus proved to be more effective than a general-purpose corpus as a pedagogical tool for the training of translation trainees. In her article, Sánchez Ramos (2020) reports on her teaching experience at the University of Alcalá (Madrid, Spain), where she taught medical translation to a group of postgraduate students. She shows the effectiveness of corpus-based approaches in improving thematic, terminological, and phraseological knowledge.

In the field of legal language and legal translation training, Biel (2010) recommends the use of a corpus-based approach along with other theoretical and practical applications. She finds that the use of parallel and translation corpora can "improve the naturalness of translation by minimising the effects of translation universals and SL interference" (Biel 2010: 13). In her opinion, monolingual and comparable corpora, on the one hand, and parallel corpora, on the other, play different roles in the training of prospective translators. The former help trainees become aware of the conventions of legal genres in the TL and the writing skills that a lawyer is required to acquire. The latter shed more light on the translation process itself and the translation techniques that a prospective translator of legal texts must possess. In his article Bertozzi (2018) presents Anglitrand, an intermodal Italian-Spanish corpus of Italian speeches delivered at the European Parliament plenary sitting in the year 2011 and their interpreted and translated Spanish version, compiled for the training of interpreters and translators.

Prieto Ramos (2019) supports the use of corpus-based approaches for quantitative and qualitative analyses in the field of Legal Translation Studies. An important requirement is that "corpus designers in this area must contextualise specific genres in their jurisdictions and branches of law, and determine their connections through inter- or intra-systemic translation or co-drafting, whether at national or international level" (Prieto Ramos 2019: 8). According to Giampieri (2021), trainees' errors of misunderstanding and mistranslation in the field of legal translation result not so much from their poor or lack of technical knowledge in the legal field as from their poor dictionary searches and their failure to consult corpora as well as analyse KWIC and collocational patterns. In other words, although subject-specific knowledge is relevant in the translation training classroom, it is also crucially important to improve the students' corpus analysis skills.

Laursen and Pellón (2012) demonstrate how corpora can improve students' skills in economic and financial translation. Relying on their teaching experience with a group of students attending a course in the translation of annual reports between Spanish and Danish, they conclude that teaching students to compile and query bilingual comparable corpora as well as training them to use concordancers and other software suites can improve their translation competence in specialised domains at terminological, genre, and stylistic levels. In addition to Laursen and Pellón's study, it is worth mentioning other corpus-informed translation training courses that rely on Do-It-Yourself (DIY) corpora. Starting from his experience with MA students in Specialised Translation at Cologne University of Applied Sciences, Germany, Krüger (2012), for instance, illustrates how DIY corpora and the Internet used as a corpus can be exploited by students for translation purposes. Like Kübler, Krüger urges trainers to teach students the theoretical rationale underlying corpora, including how to design and compile corpora using WebCorp Live. Frankberg-Garcia's (2015) MA course in translation includes the use of WebBootCat by students who crawl the web in order to compile DIY specialised corpora that they can use for their translations. In Rodríguez's (2016) research project on translation teaching via corpora, students were trained to compile DIY monolingual corpora in English and Spanish as well as parallel corpora with the help of various online platforms, e.g. Sketch Engine, WebCorp, BNC, and GloWbE. The project experience underscores the central role of learners in the translation training process.

Building DIY corpora is supposed to raise the students' cognitive skills, as Bernardini and Castagnoli (2008) contend in their review article on the role of corpora in translation teaching and practice. They suggest the adoption of an educational rather than a training attitude, in the sense that much more importance should be given to raising awareness about the use of corpora in the translation process. In the same vein, Velasco and Antonio (2013) analyses the important contribution that a corpus-based approach can give, in terms of communicative and cognitive processes, to the multimodal analysis of specialised competence. According to Liang (2020), corpora are resourceful tools for tackling the limits of dictionaries, which are decontextualised, misleading, or outdated source of information. In order to improve translation quality and inverse translation skills in a group of Chinese undergraduate students attending a translation classroom, he uses corpora as a priming tools, i.e. an "experimental paradigm for exploring the cognitive aspects of language learning and use, increasingly popular in applied linguistics studies" (Liang 2020: 217).

5.3 *Corpus-Informed Translation Training: A Case Study*

In 2016, Falco (2017) carried out a translation teaching project with a group of students attending an MA course in specialised translation, in particular the translation of legal and economic texts. Students were trained to use WebBootCat to compile DIY corpora on agreements in English and Italian. Not only did they obtain lists of terms, but they also analysed the collocational patterns associated with these terms, as the samples for "distribution agreement" (Fig. 6) and its Italian equivalent "contratto di distribuzione" show (Fig. 7).

The next step involved using the information they had collected about the grammatical and lexical-semantic behaviour of selected terms to build parallel cognitive maps both in English and Italian (Figs. 8 and 9), and identify other related terms and concepts (Fig. 10) according to a centrifugal process aimed at increasing their

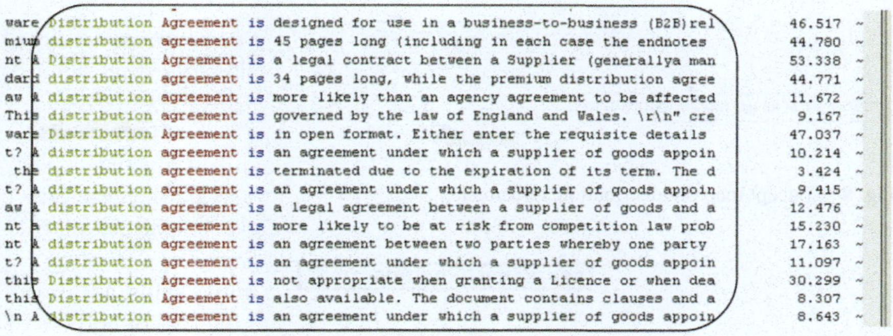

Fig. 6 Sampled concordances of "distribution agreement"

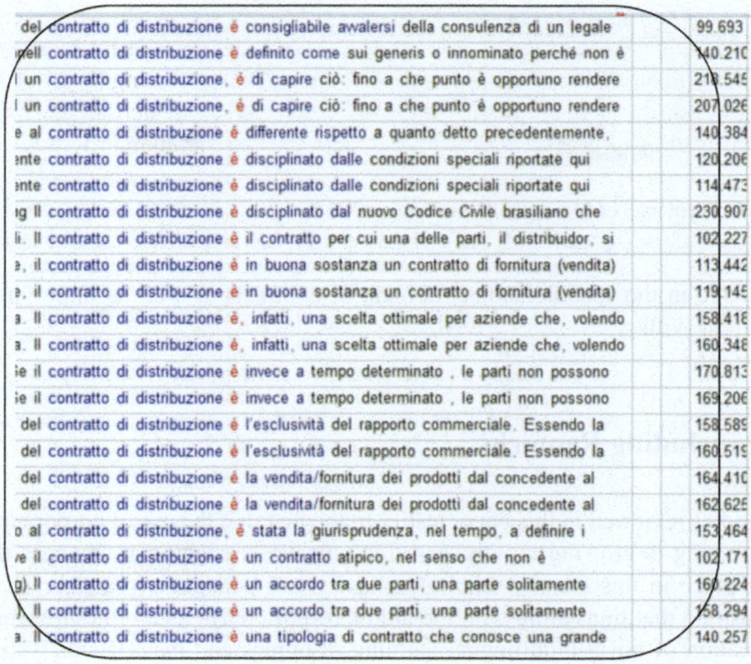

Fig. 7 Sampled concordances of "contratto di distribuzione"

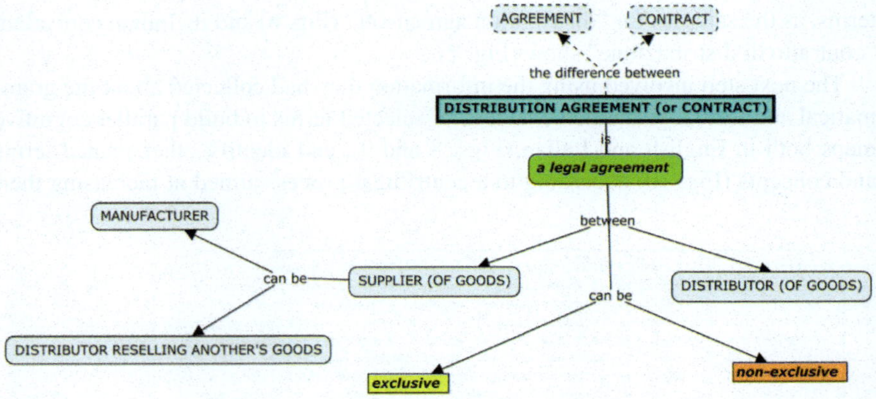

Fig. 8 Concept map of "distribution agreement"

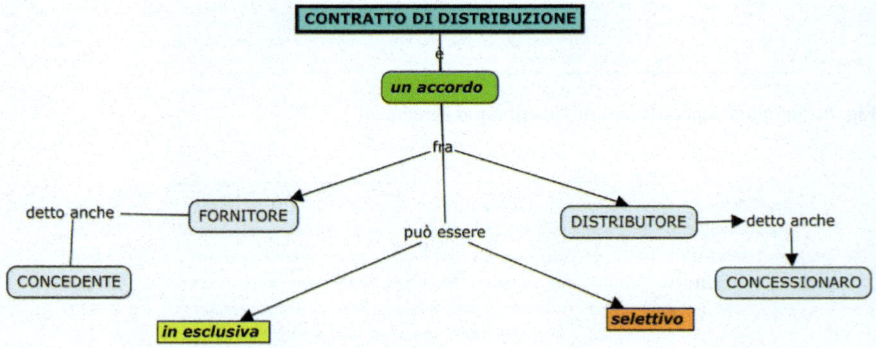

Fig. 9 Concept map of "contratto di distribuzione"

knowledge in the specific subject-domain of contracts as well as improving their translation skills.

6 Concluding Remarks

The studies reviewed in this paper testify to the fact that the use of corpora in translator training is growing rapidly, and this reflects the significant changes that are taking place in this area of scholarly enquiry and pedagogic practice. They also demonstrate that translating with the aid of corpora plays a key role in stimulating the creation of novel multilingual learning resources and materials as well as the design of new teaching procedures and testing techniques in translation education, given the growing impact of technology on present day electronically mediated communication, the study of languages, and the language industry at large.

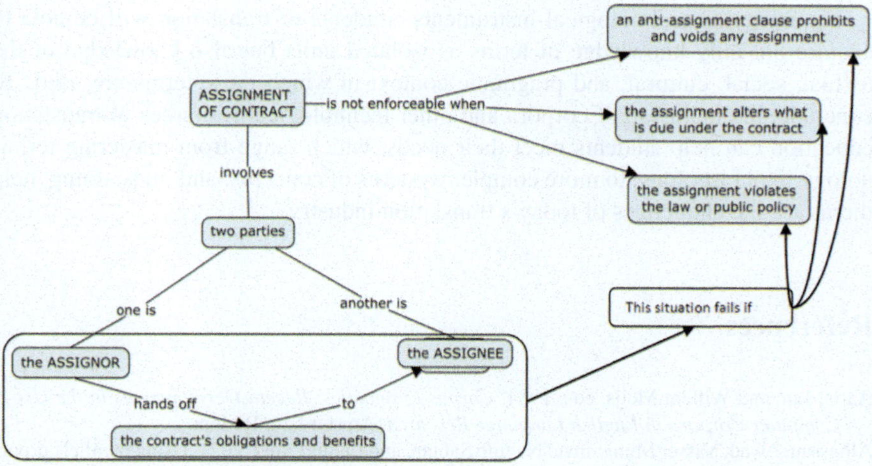

Fig. 10 Concept map of "assignment of contract"

Despite this upward trend in translation pedagogy, more empirical research is still needed in order to assess the benefits of corpus-informed translation for language as well as translation learning. In this regard, it is crucial to promote closer cooperation between educational linguists and translation studies scholars (see Laviosa and González-Davies 2020). Moreover, the incorporation of new Information Technologies and corpora in translator training, combined with student-initiated approaches in translation pedagogy, is crucially important for the development of critical thinking and autonomy as well as syllabus design and planning.

Therefore, on the one hand, what we need to do now is to boost the cognitive shift in corpus-based and corpus-driven translation teaching. Looking to the future, recent research indicates that corpus-based approaches to translator training can be further improved by incorporating other methodologies and interfacing with other areas of linguistics. By way of example, the integration of concept maps into corpus-driven teaching methods can contribute to enhancing the trainees' cognitive processes, boosting their creativity, and awareness of specialised domains, thus enabling them to acquire thematic knowledge and, consequently, perform translation tasks successfully. As Symseridou observes, "the adoption of a corpus-based teaching methodology allows for the inclusion of more specialised texts in the curriculum, even if the teacher is not acquainted with a discipline, as well as the creation of a collaborative learning environment" (Symseridou 2018: 73).

On the other hand, the fast-growing demand of technological skills in the translation industry requires the research and design of advanced modules that can contribute to enhancing the technological competence of prospective translators. To achieve these goals, research, as recommended by the EMT's *Competence Framework 2017*, will be centred on the development of search engines, translation memories, corpus-based tools, and CAT tools that can increase the trainees' impact on the translation process.

By mastering technological instruments, students of translation will be able to acquire not only knowledge of terms as isolated units but also knowledge of the textual, social, cultural, and pragmatic context in which these terms are used. To conclude, the inclusion of corpora and other technological resources in translation education can help students meet their needs, which range from mastering terminology to phraseology, to more complex systems of concepts, and, in so doing, help them face the challenges of today's translation industry.

References

Aarts, Jan, and Willem Meijs, eds. 1984. *Corpus Linguistics: Recent Developments in the Use of Computer Corpora in English Language Research.* Amsterdam: Rodopi.

Alhassan, Awad, Yasser Muhammad Naguib Sabtan, and Lamis Omar. 2021. Using parallel corpora in the translation classroom: Moving towards a corpus-driven pedagogy for Omani translation major students. *Arab World English Journal* 12 (1): 40–58.

Atan, Suryani. 2012. Towards a collaborative learning environment through ICT: A case study. In *Conference Proceedings. ICT for Language Learning. 5th Conference Edition.* https://conference.pixel-online.net/conferences/ICT4LL2012/common/download/Paper_pdf/120-IBT25-FP-Atan-ICT2012.pdf.

Baker, Mona. 1993. Corpus linguistics and translation studies: Implications and applications. In *Text and Technology: In Honour of John Sinclair*, ed. Mona Baker, Gill Francis, and Elena Tognini-Bonelli, 233–250. Amsterdam: John Benjamins.

Baker, Mon, Gill Francis, and Elena Tognini-Bonelli, eds. 1993. *Text and Technology: In Honour of John Sinclair*. Amsterdam: John Benjamins.

Beeby, Alison, Patricia Rodríguez-Inés, and Pilar Sánchez-Gijón. 2009. Introduction. In *Corpus Use and Translating*, ed. Alison Beeby, Patricia Rodríguez Inés, and Pilar Sánchez-Gijón, 1–8. Amsterdam: John Benjamins.

Bernardini, Silvia, and Sara Castagnoli. 2008. Corpora for translator education and translation practice. In *Topics in Language Resources for Translation and Localization*, ed. Elia Yuste Rodrigo, 39–55. Amsterdam: John Benjamins.

Bertozzi, Michela. 2018. ANGLINTRAD: Towards a purpose specific interpreting corpus. *inTRAlinea.* Special Issue: New findings in corpus-based interpreting studies. http://www.intralinea.org/specials/article/2317.

Biel, Łucja. 2010. Corpus-based studies of legal language for translation purposes: Methodological and practical potential. In *Reconceptualizing LSP. Online Proceedings of the XVII European LSP Symposium 2009. Aarhus 2010*, ed. Carmen Heine, and Jan Engberg. https://www.asb.dk/fileadmin/www.asb.dk/isek/biel.pdf.

Bowker, Lynne. 1998. Using specialised monolingual native-language corpora as a translation resource. *Meta* 43 (4): 631–651.

Bowker, Lynne. 2011. Off the record and on the fly: Examining the impact of corpora on terminographic practice in the context of translation. In *Corpus-Based Translation Studies: Research and Applications*, ed. Alet Kruger, Kim Wallmach, and Jeremy Munday, 211–236. London: Bloomsbury.

Colominas, Carme, and Toni Badia. 2008. The real use of corpora in teaching and research contexts. In *Topics in Language Resources for Translation and Localization*, ed. Elia Yuste Rodrigo, 71–88. Amsterdam: John Benjamins.

Falco, Gaetano. 2017. Concept maps as teaching tools for students in legal translation. *Lingue e Linguaggi* 21: 91–106.

Firth, J.R. 1957. *Papers in Linguistics 1934–1951.* London: Oxford University Press.

Frankenberg-Garcia, Ana. 2015. Training translators to use corpora hands-on: Challenges and reactions by a group of thirteen students at a UK university. *Corpora* 10 (3): 351–380.

Frérot, Cécile. 2016. Corpora and corpus technology for translation purposes in professional and academic environments. Major achievements and new Perspectives. *Cadernos de Tradução. Edição Especial: Corpus Use and Learning to Translate, Almost 20 Years*, vol. 1, 36–61.

Gallego-Hernández, Daniel. 2015. The use of corpora as translation resources: A study based on a survey of Spanish professional translators. *Perspectives* 23 (3): 375–391.

Gellerstam, Martin. 1986. Translationese in Swedish novels translated from English. In *Translation Studies in Scandinavia, Proceedings from the Scandinavian Symposium on Translation Theory (SSOTT) II Lund 14–15 June 1985, Lund Studies in English*, ed. Lars Wollin, and Hans Lindquist, 88–95. Lund: CWK Gleerup.

Giampieri, Patrizia. 2021. Can corpus consultation compensate for the lack of knowledge in legal translation training? *Comparative Legilinguistics* 46: 5–35.

González-Davies, Maria. 2004. *Multiple Voices in the Translation Classroom. Activities, Tasks and Projects*. Amsterdam: John Benjamins.

Kiraly, Don. 2000. *A Social Constructivist Approach to Translator Education: Empowerment from Theory to Practice*. Manchester: St. Jerome Publishing.

Kruger, Alet. 2004. Editorial. Corpus-based translation research comes to Africa. *Language Matters. Studies in the Languages of Africa (Special Issue: Corpus-Based Translation Studies: Research and Applications)* 35 (1): 1–5.

Kruger, Alet, Kim Wallmach, and Jeremy Munday, eds. 2011. *Corpus-Based Translation Studies: Research and Applications*. London: Bloomsbury.

Krüger, Ralph. 2012. Working with corpora in the translation classroom. *Studies in Second Language Learning and Teaching* 2 (4): 505–525.

Kübler, Natalie. 2011. Working with different corpora in translation teaching. In *New Trends in Corpora and Language Learning*, ed. Ana Frankenberg-Garcia, Lynne Flowerdew, and Guy Aston, 62–80. London: Continuum.

Laursen, Anne Lise, and Ismael Arinas Pellón. 2012. Text corpora in translator training. *The Interpreter and Translator Trainer* 6 (1): 45–70.

Laviosa, Sara (ed.). 1998. *L'Approchee Basée sur le Corpus/The Corpus-Based Approach*. Special Issue of *Meta* 43 (4).

Laviosa, Sara. 2002. *Corpus-Based Translation Studies: Theory, Findings, Applications*. Amsterdam: Rodopi/Leiden: Brill.

Laviosa-Braithwaite, Sara. 1996. *The English Comparable Corpus (ECC): A Resource and a Methodology for the Empirical Study of Translation*. PhD thesis. Centre for Translation and Intercultural Studies. University of Manchester, UK.

Laviosa, Sara, and Maria González-Davies (eds.). 2020. *The Routledge Handbook of Translation and Education*. London: Routledge.

Lee, John, Benjamin Tsou, and Tianyuan Cai. 2020. Using bilingual patents for translation training. In *Proceedings of the 28th International Conference on Computational Linguistics, Barcelona, Spain (Online), December 8–13, 2020*. 3461–3466. https://doi.org/10.18653/v1/2020.coling-main.309.

Leech, Jeffrey. 1992. Corpora and theories of linguistic performance. In *Directions in Corpus Linguistics, Proceedings of Nobel Symposium 82, Stockholm, 4–8 August 1991*, ed. Jan Svartvik, 105–122. Berlin: Mouton de Gruyter.

Liang, Biying. 2020. Corpus-based priming for inverse translation training. *Linguistics and Literature Studies* 8 (4): 215–222.

Lindquist, Han. 1989. *English Adverbials in Translation: A Corpus Study of Swedish Renderings*. Lund Studies in English 80. Lund: Lund University Press.

López Rodríguez, Clara Inés. 2016. Using corpora in scientific and technical translation training: Resources to identify conventionality and promote creativity. *Cademos de Traduçao, Florianópolis* 36 (1): 88–120. https://www.researchgate.net/publication/302969020_Using_corpora_in_scientific_andtechnical_translation_training_resources_to_identify_conventionality_and_promote_creativity.

Luz, Saturnino. 2011. Web-based corpus software. In *Corpus-Based Translation Studies: Research and Applications*, ed. Alet Kruger, Kim Wallmach, and Jeremy Munday, 124–149. London: Bloomsbury.

Marczak, Mariusz. 2016. Developing selected aspects of translation competence through telecollaboration. *English for Specific Purposes World* 16 (48): 1–12. http://esp-world.info/Articles_48/Marczak_article.pdf.

Molés-Cases, Teresa, and Ulrike Oster. 2015. Webquests in translator training. Introducing corpus-based tasks. In *Multiple Affordances of Language Corpora for Data-Driven Learning*, ed. Agnieszk Leńko-Szymańska, and Alex Boulton, 199–224. Amsterdam: John Benjamins.

Neshkovska, Silvana. 2019. The role of electronic corpora in translation training. *Studies in Linguistics, Culture and FLT* 7: 48–58.

Olohan, Maeve. 2004. *Introducing Corpora in Translation Studies*. London: Routledge.

Prieto Ramos, Fernando. 2019. The use of corpora in legal and institutional translation studies. Directions and applications. *Translation Spaces. Special Issue: Corpus-Based Research in Legal and Institutional Translation.* 8 (1): 1–11.

Velasco, Prieto, and Juan Antonio. 2013. A corpus-based approach to the multimodal analysis of specialised knowledge. *Language Resources and Evaluation* 47: 399–423.

Russo, Mariachiara, Claudio Bendazzoli, and Bart Defrancq (eds.). 2018. *Making Way in Corpus-Based Interpreting Studies*. Singapore: Springer.

Sánchez Ramos, María del Mar. 2020. Teaching English for medical translation: A corpus-based approach. *Iranian Journal of Language Teaching Research* 8 (2): 25–40

Sessoms, Diallo. 2008. Interactive instruction: Creating interactive learning environments through tomorrow's teachers. *International Journal of Technology in Teaching and Learning* 4(2): 86–96.

Shlesinger, Miriam. 1998. Corpus-based interpreting studies as an offshoot of Corpus-based translation studies. *Meta* 43 (4): 486–493.

Sikora, Iwona. 2014. The need for CAT training within translator training programmes. *inTRAlinea.* Special Issue: Challenges in translation pedagogy.

Sinclair, John. 1991. *Corpus, Concordance, Collocation.* Oxford: Oxford University Press.

Sinclair, John. 2005. Corpus and text—Basic principles. In *Developing Linguistic Corpora: A Guide to Good Practice*, ed. Martin Wynne, 1–16. Oxford: Oxbow Books.

Singer, Néstor. 2016. A proposal for language teaching in translator training programmes using data-driven learning in a task-based approach. *International Journal of English Language and Translation Studies* 4 (2): 155–167.

Symseridou, Elina. 2018. The Web as a corpus and for building corpora in the teaching of specialised translation: The example of texts in healthcare. *Fitispos International Journal* 5 (1): 60–82.

Torres-Simón, Ester, and Anthony Pym. 2019. European Masters in translation: A comparative study. In *The Evolving Curriculum in Interpreter and Translator Education: Stakeholder Perspectives and Voices*, ed. David B. Sawyer, Frank Austermühl, and Vanessa Enríquez Raído, 75–97. Amsterdam: John Benjamins.

Toudic, Daniel, and Alexandra Krause. 2017. *(On Behalf of the EMT Board) European Master's in Translation Competence Framework 2017*. Available at https://commission.europa.eu/system/files/2018-02/emt_competence_fwk_2017_en_web.pdf. Accessed May 24, 2023.

Tymoczko, Maria. 1998. Computerised corpora and the future of translation studies. *Meta* 43 (4): 652–659.

Vigier-Moreno, Francisco J. 2019. Corpus-assisted translation of specialised texts into the L2 From the classroom to professional practice. *Trans-kom* 12 (1): 90–106.

Wang, Qing. 2011. Corpus-driven learning in collegiate translation course. *Theory and Practice in Language Studies* 1 (3): 287–291.

Zanettin, Federico. 1998. Bilingual comparable corpora and the training of translators. *Meta* 43 (4): 616–630.

Zanettin, Federico. 2011. Hardwiring corpus-based translation studies. In *Corpus-Based Translation Studies: Research and Applications*, ed. Alet Kruger, Kim Wallmach, and Jeremy Munday, 103–123. London: Bloomsbury.

Zanettin, Federico, Silvia Bernardini, and Dominic Stewart (eds.). 2003. *Corpora in Translator Education.* Manchester: St. Jerome Publishing.

Sara Laviosa is Associate Professor of English at the University of Bari 'Aldo Moro', Italy. She has published extensively in international journals and collected volumes, and is author of *Corpus-Based Translation Studies* (Rodopi/Brill, 2002) and *Translation and Language Education* (Routledge, 2014). She is co-author (with A. Pagano, H. Kempannen and M. Ji) of *Textual and Contextual Analysis in Empirical Translation Studies* (Springer, 2017). Her recent publications include *The Routledge Handbook of Translation and Education* (co-edited with M. González Davies, 2020), *The Oxford Handbook of Translation and Social Practices* (co-edited with M. Ji, 2020), *CTS Spring-Cleaning: A Critical Reflection, Special Issue of MonTI* (co-edited with M. Calzada Peréz, 2021), and *Recent Trends in Corpus-based Translation Studies*, Special Issue of *Translation Quarterly* (co-edited with Kanglong Liu, 2021).

Gaetano Falco is Associate Professor of English at the University of Bari 'Aldo Moro', Italy. His main research interests include Translation Studies, Translation Teaching, Translation of LSPs, Critical Discourse Analysis, Cognitive Linguistics and Corpus Linguistics. In 2014, he published his monograph, *Metodi e strumenti per l'analisi linguistica dei testi economici. Dalla SFG al Web 2.0* (Bari: Edizioni dal Sud). He has also published journal articles and book chapters on translation teaching, translation of economic discourse in professional and non-professional genres (e.g. academic journals, comic books, movies), and CDA-based studies on corporate discourse.

Corpora, Machine Learning and Post-editing

Applying Incremental Learning to Post-editing Systems: Towards Online Adaptation for Automatic Post-editing Models

Marie Escribe and Ruslan Mitkov

1 Introduction

Machine Translation (MT) emerged in 1949, when Warren Weaver, a researcher at the Rockefeller Foundation, proposed the use of computers for automatic translation based on information theory and successes in code breaking during the Second World War (Hutchins 2005). A few years later, in 1954, a public demonstration of an MT system (the Georgetown-IBM experiment) took place in New York (ibid.), generating great enthusiasm among the research community despite the limited performance of this system. However, MT faced harsh criticism, with one of the earliest examples being the Automatic Language Processing Advisory Committee (ALPAC) report in 1966, which concluded that MT outputs were too disappointing to continue investigating such systems, especially since there were sufficient translators to complete translation projects (ibid.).

Today, this conclusion no longer stands: translation plays a crucial role in enabling international communication and translators endeavour to deliver high volumes in record times and often rely on technological assistance, including Computer-Assisted Translation (CAT) tools and MT systems. With the recent advances in deep learning, Neural MT (NMT) reached unprecedented quality levels, thus leading to the integration of MT technologies in the modern translation workflow. Consequently, the role of translators is progressively shifting towards post-editing (PE). Nevertheless, although PE can boost productivity, it can be more demanding than translation from scratch. Re-training MT engines is a possible option to improve their performance, but it may not always be possible as this requires having access to certain MT system

M. Escribe (✉)
Universitat Politècnica de València, Valencia, Spain
e-mail: mcescrib@doctor.upv.es

R. Mitkov
Lancaster University, Lancaster, England
e-mail: r.mitkov@lancaster.ac.uk

J. Pan and S. Laviosa (eds.), *Corpora and Translation Education*, New Frontiers in Translation Studies, https://doi.org/10.1007/978-981-99-6589-2_3

parameters. Automatic post-editing (APE) models, in contrast, do not need such information, as they are trained to detect and correct errors typically found in MT outputs based on past observations.

Nevertheless, APE has not benefited from interaction with the translator and only a few attempts have been made to study the online learning algorithms behind adaptive APE models. This study aims to address this gap by implementing and examining online adaptations of APE models and comparing these with models trained in batch mode. For further analysis, this comparison also encompasses several language pairs and domains.

2 Related Work

2.1 Post-editing and Latest Machine Translation Systems

The first MT systems relied on a limited set of linguistic rules to produce a transitional representation used to generate a translation. In the 1980s and 1990s, MT research focussed on corpus-based techniques, leading to the creation of example-based MT, and later to statistical MT (SMT). The idea of using artificial Neural Networks (NNs) for MT was formulated by Ñeco and Forcada (1997) and Castaño and Casacuberta (1997) shortly after but could not be implemented due to the limited computing power available at the time. Recently, however, technological advances allowed for developing NMT, which is now considered the state of the art. Consequently, an increasing number of language service providers have now integrated MT solutions in their workflow and rely on linguists to perform PE (which is defined as the process of correcting an MT output).

However, despite significant gains in performance over the years, NMT outputs are far from being perfect and the recent claims of NMT reaching human parity (Hassan et al. 2018) have been strongly criticised (Toral et al. 2018; Poibeau 2022). NMT is indeed prone to certain errors, such as omissions, additions and mistranslations (Castilho et al. 2017). Hallucinations are also frequent (Guerreiro et al. 2022) and constitute a serious threat in MT outputs. Further issues are typically related to multi-word expressions (MWEs; Zaninello and Birch 2020), syntax (Castilho et al. 2017) and the document level (Castilho 2020). Such deficiencies only contribute to increasing the PE effort.

2.2 Automatic Post-editing

The aforementioned difficulties in traditional PE led to the emergence of APE, which aims at automatically correcting an MT output to improve its quality (do Carmo et al. 2020). This is achieved using machine learning (ML) algorithms which are trained to detect and correct errors based on datasets containing triplets of source–MT output–human post-edit (SRC-MTO-HPE). According to Bojar et al. (2015), one of the main objectives of APE is to improve MTOs by exploiting information unavailable to the MT decoder. APE can also serve as a domain adaptation method for adjusting the terminology and style of a text to a certain field.

APE experienced similar developments as MT: it was first based on rules, then adopted statistical methods before making use of ML and NNs. Just like MT, APE faced some discouragement from the research community on several occasions. The first APE studies emerged in the 1990s, when Knight and Chander (1994) proposed rule-based APE modules complementing MT systems to improve MTOs. These modules considered the MTO as a "source" to be "translated", and the original version of the text was not considered. Having observed that recurring errors are often found in MTOs, Allen and Hogan (2000) introduced an APE module based on a controlled language and adopted an ML approach using SRC-MTO-HPE triplets to extract PE rules in order to re-apply these to unseen texts. This approach sets the foundation for subsequent studies, as APE today still relies on triplets for training. Taking advantage of advances in SMT, Simard et al. (2007) suggested the use of an SMT system for APE, thus assuming that PE could be cast as a monolingual translation task.

The field of APE gained popularity with the emergence of the first APE shared task at WMT'15.[1] This shared task has been running ever since, providing triplets in a variety of domains and languages and became an international forum for discussing advances in APE. The first round was not very successful and was described as the 'stone age' of APE by Junczys-Dowmunt (2018), due to the low performance of the submissions, as none could beat the baseline (i.e. unedited MTO). Bojar et al. (2015) attributed this to the statistical method used, the nature of the data (news domain, which is unrestricted) and the PE procedure (performed by crowd-sourced workers, who tend to be inconsistent and to implement unnecessary changes, thus affecting homogeneity). Consequently, the 2016 round was based on a technical domain (IT) and PE was performed by professionals, which resulted in systems with higher performance (Bojar et al. 2016). This improvement was also attributable to the use of new techniques, in particular NNs. The results of the following round were even more encouraging (Bojar et al. 2017), hence making 2017 the "golden age" of APE (do Carmo et al. 2020) and confirming the efficiency of NNs. In 2018, NMT outputs were used for the first time in the shared task. While all the neural models managed to beat the baseline, working with NMT outputs proved to be challenging, given the smaller margin this leaves for improvement (Chatterjee et al. 2018). For this reason, do Carmo et al. (2020, p. 5) described this stage of APE as 'approaching its demise'. From 2019, the shared task focussed on NMT only (Chatterjee et al. 2019), which resulted in the production of unnecessary corrections (Chatterjee et al. 2019; do Carmo et al. 2020). The most recent round (Chatterjee et al. 2020) focussed

[1] WMT: Workshop on Machine Translation. While the research community continued using this acronym, WMT is now a well-established international conference series (Conference on Machine Translation).

on NMT outputs of texts belonging to a generic domain, which also constituted a significant difficulty compared to the previous rounds (which focussed on the IT domain since 2016). However, the overall outcomes were positive, as considerable improvements over the baseline were reached.

Despite the achievements made over time, notably using neural technologies such as the transformer architecture (Vaswani et al. 2017), APE is still faced with a number of challenges. In particular, it is common for APE systems to produce overcorrections and fails to detect certain errors (do Carmo et al. 2020). Moreover, human evaluators tended to attribute low scores to APE outputs in most rounds of the shared task (except in 2020). Therefore, although APE may contribute to improving MT, the target texts thus generated require to be proofread by a human expert (Shterionov et al. 2020).

2.3 *Human–Computer Interaction in Translation Technologies*

Today, it is common for translation projects to be performed in a CAT environment, including translation memories (TMs), which recycle past translations and thus contribute to improving efficiency (Mitkov 2021). TMs are updated on the fly with segments validated by the translator, which constitutes valuable interaction between the CAT tool and its user. This database has the clear advantage of being tailored to the style of the translator and to the document being translated, which is particularly useful for enhancing consistency in large-scale projects.

As far as PE is concerned however, human–computer interaction (HCI) remains limited because post-editors correct MTOs without feeding any of the changes back to any database. In fact, HCI was at the centre of research in the early days of MT (it was notably advocated by Bar-Hillel 1960 and Licklider 1960, see Sect. 2.5.1), but the field progressively moved away from the HCI perspective after the 1980s to focus on fully automatic translation (Green et al. 2015).

Nevertheless, the situation has evolved since then, and several projects introduced interactive MT (IMT) systems, with one of the first IMT attempts being the study by Foster et al. (1997). The HCI approach was successfully exploited later in the TransType (TT) and TransType-2 (TT2) projects (Foster et al. 2002; Esteban et al. 2004, respectively). These systems aim to generate future word(s) (the 'suffix') dynamically after the translator enters the beginning of a translation (the 'prefix'). The suggested suffixes can then be either validated or amended by users, and the system can in turn exploit this information to make informed predictions. Consequently, IMT models must be continuously updated, which is typically achieved via online learning techniques. For instance, Ortiz-Martínez et al. (2010) introduced an interactive SMT system based on online learning, in which the feature values are updated each time a new sentence pair becomes available, and translation suggestions are made by restricting the search space to segments containing a particular prefix. Similar studies include the Multimodal Interaction in Pattern Recognition and Computer Vision project (MIPRCV; Toselli et al. 2011) and the Cognitive Analysis and Statistical Methods for Advanced Computer-Aided Translation project (CASMACAT; Alabau et al. 2013). Moreover, Ortiz-Martínez and Casacuberta (2014) introduced an updated version of the Thot toolkit. In addition to standard SMT features, it also allows for using online learning techniques to implement interactive SMT and was designed to be integrated into CASMACAT. This version of Thot was also used in further IMT systems as well as in online adaptation of APE models (see Sect. 2.5.1).

Following the neural paradigm, recent IMT projects have tended to focus on NMT. Indeed, Knowles and Koehn (2016) introduced an interactive NMT system, in which the translation prediction is conditioned on a given prefix. Santy et al. (2019) argued that IMT performed well mostly in resource-rich scenarios and therefore introduced an interactive NMT system designed for low-resource languages. Peris and Casacuberta (2019) also developed an interactive NMT system based on constrained beam search and focussed on reactivity by updating the model based on single-character interactions.

2.4 *The Impact of Interactive Translation Tools on the PE Effort*

HCI has proven highly beneficial, as it contributes to significantly reducing the cognitive effort involved (Alves et al. 2016b). This claim should nevertheless be toned down, as some studies did not find significant improvements in interactive settings. This was the case for Underwood et al. (2014), who reported mixed results for CASMACAT, with certain participants not finding the interactive setting helpful, while others reported more positive experiences. Alves et al. (2016a) found that IMT did not yield improvements in efficiency but could contribute to reducing the cognitive effort. Several explanations for these results can be suggested, such as the lack of familiarity of the participants with such interfaces and the time and effort required to engage with predictions which are constantly being updated. More recently, Karimova et al. (2018) performed a user study comparing the PE effort in the case of traditional and adaptive NMT models. The outcomes revealed a significant effort reduction in the online scenario. Similarly, Domingo et al. (2019) conducted a user study to assess incremental adaptation of NMT and concluded that such a technique improved productivity and quality. Domingo et al. (2020) extended this study and reached similar conclusions. In particular, users noticed that the adaptive systems were able to efficiently exploit inputted post-edits and make predictions tailored to the text at hand, which contributed to reducing the effort involved. Overall, the results reported in these studies indicate that the user experience can be enhanced in interactive settings, which reinforces the need for further research on interactive MT and PE.

2.5 *Towards Interactive Translation and Post-editing Environments*

2.5.1 Incremental Adaptation of PE Models

In the case of APE, the system provides an output which is subsequently corrected by humans, which impedes any form of HCI. Nevertheless, human input in an interactive environment provides a unique opportunity to improve the PE process (Escribe and Mitkov 2021). Moreover, given that APE outputs require editing, focussing on assisting post-editors rather than seeking to make the PE process fully automatic would be highly beneficial. In fact, the use of PE information to improve MTOs has been explored in previous studies. Nishida et al. (1988) were among the first authors to suggest using a feedback system based on PE information to improve an MT engine. Similarly, Su et al. (1995) used feedback to adjust the parameters of an MT system to comply with preferred stylistic standards, and Phaholphinyo et al. (2005) introduced a method to improve an RBMT system using PE information.

Despite successfully using PE feedback, these approaches aimed at improving a particular MT system. While learning from PE enabled directly fixing the MT system, it can be argued that a post-processing of the MTO would also yield good results. This method is reminiscent of APE models, as such systems do not require any access to MT system parameters to perform corrections.

In fact, while HCI has been explored using glass-box approaches in IMT settings, several studies have also introduced "interactive" PE models. Like APE, interactive PE is a black-box approach, as it does not require access to the MT system parameters, but only to an MTO. An interactive PE model would therefore learn from corrections as PE is being carried out. This concept was formulated as early as 1960 by Bar-Hillel (who advocated for a "machine-post-editor partnership") and Licklider (who described an 'anticipated symbiotic partnership' between human and machines). This approach is in line with the findings of the user survey conducted by Lagoudaki (2008), which revealed that translators deemed it important for a system to adapt to the user input in order to reduce errors in future translations.

The study by Knight and Chander (1994, p. 779) was the first to implement an "adaptive post-editor": a module exploiting corrections on the fly to learn "to emulate what the human is doing". Simard and Foster (2013) introduced Post-Edit Propagation (PEPr), a model which takes an SMT output and uses an APE module based on online methods to learn from human corrections on the fly. This work is largely inspired by how TMs recycle previous translations, as the assumption is that automatically propagating corrections can be beneficial when the number of repetitions in a text is high. Building on this model, Lagarda et al. (2015) used the Thot toolkit (Ortiz-Martínez and Casacuberta 2014) to design an online APE system specifically designed for domain adaptation via an automatic correction of repeated errors. The reported results showed that both online models (Simard and Foster 2013; Lagarda et al. 2015) were able to beat the baseline in the case of a high repetition rate.

2.5.2 Online APE

More recently, Chatterjee et al. (2017b) emphasised the need for APE systems capable of handling continuous streams of data to adapt to evolving settings and the variety of domains in real-world translation workflows (which they refer to as a "Multi-Domain Translation Environment" or MDTE). Indeed, PEPr (Simard and Foster 2013) is based on the document level and its parameters are reset when coming to a new document, which impedes reusing relevant knowledge already acquired. Based on this observation, Chatterjee et al. (2017b) introduced APE components responsible for retaining and selecting relevant information. In this model, a selection technique (called "instance selection mechanism") is used to identify relevant rules to apply to the current segment (according to a similarity score based on TF-IDF[2]) and the edit operations performed are incrementally added to an index (the "dynamic knowledge base"). This information is then exploited at decoding time to provide the decoder with a selection of only the best options, based on the edits previously observed. The results revealed that significant performance gains were achieved with the online models. It should be noted that this approach is reminiscent of the work of Farajian et al. (2017), who sought to adapt NMT on the fly in an MDTE scenario. The proposed model is based on an attentional encoder-decoder architecture (Bahdanau et al. 2014), with both the encoder and the decoder implemented as Gated Recurrent Units (GRUs; Cho et al. 2014). In this model, an instance selection step is also employed based on a retrieval mechanism which selects the parallel sentences with the highest similarity scores (with respect to the current segment) and uses the retrieved pairs dynamically to produce a translation and update the model parameters. The results revealed that such a model could outperform traditional NMT and domain-adapted NMT models in an MDTE setting.

Building on the work of Chatterjee et al. (2017b), Negri et al. (2018a) developed an APE system which can learn from simulated interaction with post-editors. To achieve this, they used a synthetic corpus to train their system (Synthetic Corpus for Automatic Post-Editing [eSCAPE], Negri et al. 2018b). In the context of APE, a synthetic corpus refers to a bilingual parallel corpus (containing a source text and a human-translated version) to which MT is applied to the source side in order to generate an MTO version. Consequently, such a corpus artificially reproduces a PE scenario, as the original (human translated) target text is considered as the post-edited version of the MTO side. The authors also adopted an online approach, with each new correction being passed to the system in a continuous learning mode. In practice, similar components to those described by Chatterjee et al. (2017b) are applied to make relevant selections (instance selection mechanism and knowledge base). The authors reported that this online approach outperformed generic APE systems for both general and specialised domains, which makes this method promising for future

[2] TF-IDF: Term Frequency-Inverse Document Frequency. This measure was proposed by Salton, et al. (1975) in their theory of term importance, which demonstrated that the relevance of a word is subject not only to its frequency but also its specificity in a document.

developments. It also demonstrates that synergy between an APE system and its users (although simulated) can yield better results compared to fully automatic PE.

2.5.3 Batch Versus Online Learning for APE

Most of the work implementing interactive models has primarily focussed on MT (glass-box IMT systems) and only a few studies have explored this solution for PE (black-box PE systems). However, the underlying assumption is similar: in both cases, the model learns an adaptation behaviour. Despite the recent interest in developing interactive systems, APE research focusses mainly on improving MTOs without implementing any form of HCI. Most APE systems operate in batch training: the corpus is entirely available at once, and the model learns in an offline mode. However, online learning has several advantages over batch training. In particular, since online algorithms take one data point at a time, they can process continuous data streams, and thus do not require retraining when new data becomes available. Moreover, online models are sensitive to concept drift, which is particularly relevant in the MDTE hypothesis. The primary downside of online algorithms, however, is catastrophic forgetting, a circumstance in which online models forget previously learned knowledge as new information becomes available.

Chatterjee et al. (2017b) and Negri et al. (2018a) explored online adaptation of APE systems, and both studies simulated the interaction with the translator using reference translations. This choice is understandable, as creating fully interactive environments is a complex task, and deploying human-in-the-loop approaches requires a wide range of skills, including designing an interface to retrieve PE input. The clear advantage of interactive systems over APE is that they do not require access to a PE reference beforehand, since it is continuously inputted by the user. This is a significant benefit given the scarcity of data (triplets). Nonetheless, creating interactive environments is not always feasible, mostly for reasons of time and cost, which justifies the choice of simulating the interaction with the translator. Nevertheless, simulating human interactions has numerous implications, with the most salient being that corrections are limited to pre-existing references, as it limits the corrections to match one gold standard. This is a rather unrealistic scenario, as translation is an open-ended problem. Consequently, in the case of APE, such approaches are likely to affect the performance by generating overcorrections (Domingo et al. 2019).

3 Methodology

3.1 Motivation and Research Questions

With the increasingly high quality of NMT outputs, PE is becoming the norm of the modern translation workflow. However, PE can be challenging, especially because it lacks HCI. While IMT models have been introduced to address this issue, such systems require having access to the MT system parameters, which might not always be possible. APE models, in contrast, provide corrections of an MTO in a black-box scenario but they do not benefit from interaction with the translator. This study seeks to bridge this gap by proposing online APE models and comparing their outputs to similar systems trained in a batch learning mode. Furthermore, while synthetic data is typically used in APE research, this project avoids using synthetic corpora.

It should be noted here that developing a fully interactive PE model (in which corrections are performed in real time by a translator) raises several challenges, in particular designing a user interface and relying on volunteers to dynamically post-edit translations to train the underlying model. Consequently, while this option remains highly relevant, it was deemed to fall beyond the scope of this project. Although an interactive functionality is designed as a proof of concept, the interaction with the translator is therefore simulated using post-edited versions of MTOs.

Chatterjee et al. (2015, p. 157) formulated three questions which should be considered when designing an APE project:

- *Q1: Does APE yield consistent MT quality improvements across different language pairs?*
- *Q2: What is the relation between thee original MT output quality and the APE results?*
- *Q3: Which of the […] analysed APE methods has the highest potential?*

Taking these research questions into consideration, the primary objective of this study is thus to examine the performance of online APE models in different settings (comparing different language pairs, domains, types of MT systems). Two additional research questions are also explored:

- *Q4: Is online adaptation an efficient strategy to improve the performance of APE?*
- *Q5: How do APE models trained on non-synthetic data perform?*

Consequently, the aim of this project is not directly to improve traditional APE systems but rather to explore online adaptations of APE systems and compare them with batch models.

3.2 Data Selection and Processing

3.2.1 Selection

Data availability is one of the main challenges in APE research. In fact, the corpora made available for the APE shared tasks constitute the largest source of data. Another alternative consists of generating triplets artificially, following the methodology adopted by Negri et al. (2018b) to compile eSCAPE (see Sect. 2.5.2). Such an approach appears relevant to create large corpora, which are needed when working with NNs. However, relying on synthetic data might not be optimal, as pre-existing translations may be more different to the MTO than post-edits, which can make the system learn overcorrections. Furthermore, while incremental adaptation is not a novel research avenue in translation technologies, only two recent studies have applied this method to APE: Chatterjee et al. (2017b) and Negri et al. (2018a). Chatterjee et al. (2017b) worked in EN-DE and used the APE'16 data as well as a subset of the Autodesk corpus (Zhechev 2012), in which the MTO side was produced by MT engines older than NMT. Negri et al.'s study (2018a) focussed on EN-IT and utilised a synthetic corpus (eSCAPE), in which the MTO side was obtained using both PBSMT and NMT. Based on these observations, the primary criteria for corpus selection were the following:

- Using non-synthetic data
- Using a corpus in which the MTO side was generated by an NMT system
- Using also a corpus in which the MTO side was generated by an older MT system
- Testing both in-domain and generic corpora
- Experimenting with different language pairs

In addition to the aforementioned criteria, preferences in language combinations were also considered, which resulted in the selection of three corpora in three different language pairs: English to Spanish (EN-ES), German to English (DE-EN) and English to Mandarin Chinese (EN-ZH). The basic features of these datasets are summarised in Table 1.

Table 1 Features of the corpora used in this study

	APE'15	APE'17	APE'21
Language pair	EN-ES	DE-EN	EN-ZH
Domain	News	Pharmacological	Wikipedia
Number of triplets (train + dev)	12,272	26,000	8,000
MT system used	Unknown	SMT	NMT
HPE	Crowdsourcing	Professionals	Professionals

3.2.2 Processing

The original training sets were split into train and test sets (the latter taking between 10 and 15% of the original training sets for testing), and the development data sets were used for updating the pre-trained models in both batch and online modes.

Several pre-processing functions were created. In particular, start and end tokens were added to each segment, the data sets were cleaned by removing special characters and each segment was padded to a maximum length. Two dictionaries were also created (a word index and a reverse word index). The maximum vocabulary size was set to 15,000 for EN-ES and to 18,000 for DE-EN. Regarding EN-ZH, it was limited to 18,000 for EN and 22,000 for ZH. Several standardisation operations were applied, including Unicode normalisation in EN-ES and DE-EN. For EN-ZH, spaces were already included to separate words in the WMT dataset (although Chinese is an unsegmented language), and this information was therefore used for tokenisation and unwanted characters and symbols were removed.

3.3 Models' Design

Several models were created for each language pair. First, a batch APE model was defined and pre-trained using the training set. Since it served as a basis for updating the two other models, this first batch pre-trained model is referred to as 'BASE'.

Then, BASE was updated using the development set in batch and online modes. The model updated in batch mode is called 'BATCH', and the model updated in online mode, 'ONLINE'.

3.3.1 Batch Models

BASE receives the SRC and the MTO as input and attempts to predict the HPE side. Following recent publications suggesting that NNs are state of the art for both MT and APE, this model is based on an encoder-decoder neural architecture. The backbone of this system is inspired by the FBK's participation in the APE'17 shared task (Chatterjee et al. 2017a) which yielded the best results in the 2017 round, and the submission of Saarland University and DFKI in the 2018 round (Pal et al. 2018).

Certain approaches regard APE as a monolingual task and focus solely on target information (e.g. Pal et al. 2017). However, SRC information remains crucial, especially to disambiguate corrections (Chatterjee et al. 2017a; Pal et al. 2018). Consequently, in the present work, a bidirectional encoder takes the SRC and its corresponding MTO and encodes them separately using a GRU model. After encoding, these segments are passed to the decoder (in the form of an SRC-MTO tuple) together with a hidden state, which is used to initialise the decoder's hidden state.

The decoder is composed of two GRU models and takes the previous word in the sentence, its previous hidden state, and the encoded SRC and MTO segments as input.

The decoder predicts the next word in the HPE side after executing a series of steps. First, it concatenates the previous hidden state and the previous word embedding and passes this concatenated vector through the first GRU to obtain an intermediate hidden state. Then, it initialises two attention mechanisms (Bahdanau et al. 2014) and uses both to attend to the encoded SRC and MTO. This allows for weighting each word in the SRC and MTO using intermediate hidden states and therefore attributing a higher weight to relevant words at each time step. The attention attends to its encoded inputs separately using the intermediate hidden state as a query to produce two context vectors: $\vec{s}$ for SRC and $\vec{m}$ for MTO. These context vectors are then merged and linearly transformed using a dense layer to produce the final context vector $\vec{v}$. This vector $\vec{v}$ and the intermediate hidden state are then concatenated, and the resulting vector is passed to the second GRU to produce a final hidden state. Thereafter, the hidden state is passed to a dense layer to produce the probability of the next target word in the HPE side.

A loss class is also created to perform the calculation of the loss between the true HPE and the model's predictions based on sparse categorical cross entropy. A customised training function is defined to execute a series of operations. In particular, it passes the SRC and MTO segments to the encoder as input, and it uses the returned hidden state of the encoder to initialise the decoder's hidden state. Then, it uses a loop to iterate through each word to make predictions.

The loss is calculated using teacher forcing, which allows for continually feeding the next correct word in the real HPE sentences to the model instead of passing the previous output back to the model. This makes the model learn faster, as it is fed with the correct prediction at each step (instead of its own output, which can be incorrect), which strengthens the mapping between the previous word and the predicted word.

Finally, an optimisation process is necessary to reduce the loss. This is achieved via a gradient descent algorithm which seeks to make the model predictions converge to the lowest loss possible. Here, the training function applies an optimisation procedure using the Adam optimiser (Kingma et al. 2015).

As for hyperparameters, the embedding dimension and the number of linear units were set to 512, the number of attention units was equal to 256 and the size of the hidden dimension was set to 1024. Regarding regularisation, a dropout ratio of 0.2 was applied to avoid overfitting (Hinton et al. 2012). The Learning Rate (LR) was initialised to 0.001 and a formula was applied to reduce it for every epoch (see Sect. 3.3.2). To prevent the model from overfitting, an early stopping monitor with a patience of 4 was used.

It is important to mention here that it was not possible to use the same number of neural units as in Chatterjee et al. (2017a) due to the limited hardware resources available. Another significant difference with this work lies in the amount of data. For instance, Negri et al. (2018a) trained their model using the EN-IT section of eSCAPE (6.6 million triplets). The size of the datasets used in the present work is comparatively limited.

After training, a function performs the (SRC, MTO) $\rightarrow$ HPE transformation. To that end, the model requires access to the HPE from the mapping provided by the processing functions. Overall, this function is similar to the training loop executed by the custom training function, except that the input to the decoder at each time step is a sample from the decoder's last prediction.

After defining the batch APE model, an updated model is trained in batch mode (BATCH) using the development sets, based on the same approach used for training BASE.

3.3.2 Online Adaptation and Interactive Functionality

An online adaptation algorithm of this model (ONLINE) was also created based on the work on multi-domain adaptation for NMT by Farajian et al. (2017) as well as the online neural APE model presented by Negri et al. (2018a). This adaptation takes BASE and uses the development data to update it incrementally. It requires a model (pre-trained or untrained) and is composed of a Knowledge Base (KB) and a retrieval engine (Fig. 1). The KB either contains training data or is empty (in this case, it is used to store subsequent SRC-MTO-HPE data during the online learning process).

ONLINE takes a single SRC-MTO-HPE datapoint at a time for learning. The SRC-MTO pair is first used to query the KB using the retrieval engine. The retrieval engine takes the SRC segment and searches for similar datapoints in the KB based on the cosine similarity. To that end, the current SRC and the segments stored in the KB are converted to vectors based on TF-IDF and the cosine similarity between the SRC and datapoints stored in the KB is computed.

Thereafter, candidates are sorted in descending order based on their similarity scores. This allows for finding the datapoint(s) above a specified threshold (here,

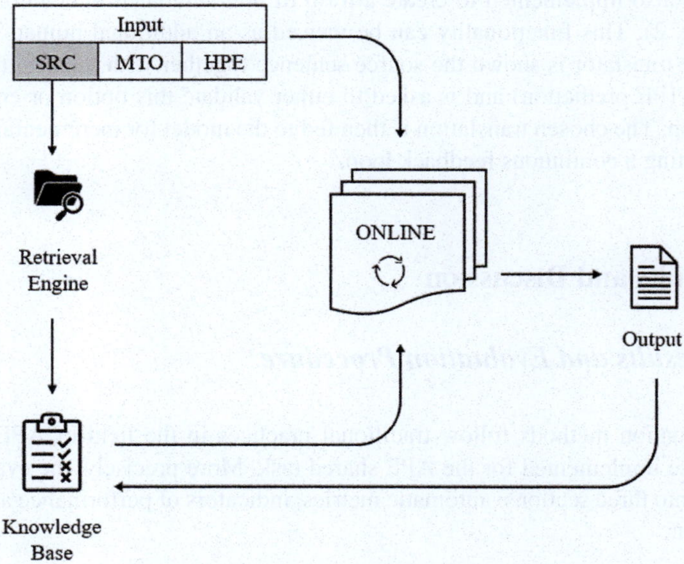

Fig. 1 Architecture of the ONLINE model

0.35). If no datapoint is found above this threshold (or if the KB is initially empty), the HPE of the SRC-MTO pair is given using the generic BASE. The newly obtained datapoints are then used to update the parameters. The updated model is then used to predict the HPE of the SRC-MTO pair. The model parameters are subsequently updated again given this single SRC-MTO and its original HPE for several epochs and the newly learned datapoint is added to the KB.

Certain limitations should be acknowledged here. In particular, the current sentence is judged to be similar to the reference sentence if it is composed of similar words, thus ignoring word order and context. Furthermore, since training is restricted to one datapoint at a time, the LR and the number of epochs play a crucial role, as excessively high/small values can render training unstable and/or inefficient. To avoid overfitting, the number of epochs is reduced to three and a time-based decay is applied to reduce the LR when increasing the number of epochs. This is because the model is likely to overlearn from the error calculated during optimisation.

It should also be mentioned that using BASE as a basis for the online adaptation enables the use of the KB already populated with training instances instead of leaving it empty. Given that the data available is limited for this project, training an online model from scratch does not seem the most appropriate method. Moreover, analysing the performance of a batch model and an online model trained from scratch does not make for a fair comparison. Indeed, an online model is inherently slower to train and would take a significantly longer time to reach convergence, compared to a traditional system. In contrast, using a pre-trained model as a basis for updating both batch and online models appears fairer.

The online model described above was implemented using the HPE as a reference, thus simulating the interaction with the translator. However, an interactive functionality was also implemented to create a form of non-simulated interaction with the user (Fig. 2). This functionality can be viewed as an additional human validation step. The translator is shown the source sentence together with the best translation available (PE prediction) and is asked to either validate this option or enter a new translation. The chosen translation is then fed to the model for incremental learning, thus creating a continuous feedback loop.

4 Results and Discussion

4.1 Results and Evaluation Procedure

The evaluation methods follow traditional practices in the field of APE, particularly those implemented for the APE shared task. More precisely, the evaluation is divided into three sections: automatic metrics, indicators of performance and human evaluation.

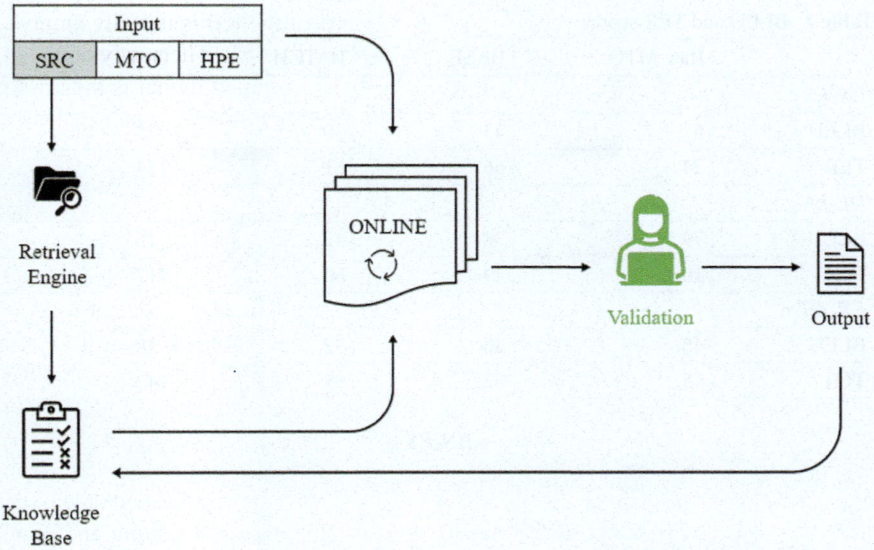

Fig. 2 Interactive functionality in the ONLINE model

4.1.1 Automatic Metrics

The models were evaluated using metrics traditionally employed for assessing MT, namely TER (Translation Edit Rate; Snover et al. 2006) and BLEU (BiLingual Evaluation Understudy; Papineni et al. 2002). These scores are computed in case-insensitive mode taking the HPE as a reference. Following the evaluation procedure of the APE shared task, the scores are compared against the 'do nothing' baseline (unedited MTO). A high TER score would suggest that a candidate APE output requires heavy editing (to be transformed into the reference HPE), whereas a low TER score indicates that the output is similar to the reference and is therefore of a higher quality. In contrast, a low BLEU score would suggest that a candidate is excessively different from the reference, and a high BLEU score indicates a high quality. These scores are reported in Table 2.

In their analysis of both batch and online APE models, Chatterjee et al. (2017b) plotted the TER moving average, a graph showing TER for each sample (i.e. test instances) of the test set in a window containing a given number of data points, thus allowing for visualising the performance of the models across the data. This measure was also used here for all models and language pairs, and the plots obtained appear in Fig. 3.

Table 2 BLEU and TER scores

	Raw MTO	BASE	BATCH	ONLINE
EN-ES				
BLEU	67	43	30	35
TER	24	66	61	134
DE-EN				
BLEU	79	38	42	16
TER	16	49	60	117
EN-ZH				
BLEU	46	40	32	28
TER	45	82	98	142

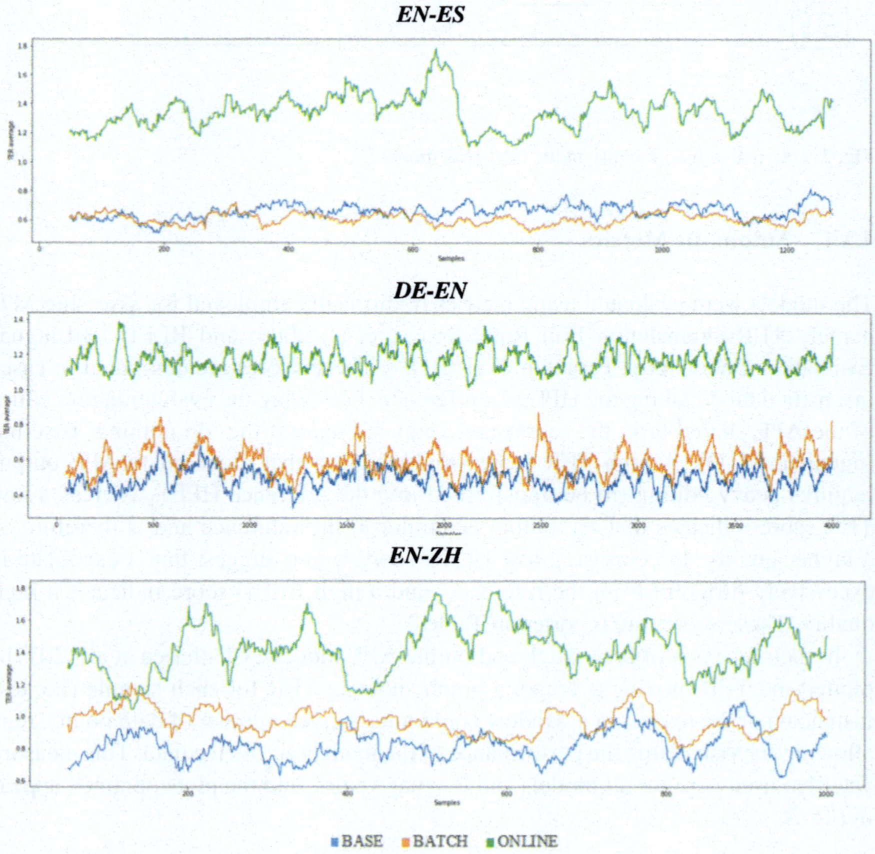

Fig. 3 TER moving averages

Table 3 Macro indicators

	BASE	BATCH	ONLINE
EN-ES			
Modified	93.47	96.07	100
Improved	7.78	2.83	0.24
Deteriorated	80.74	88.99	99.37
Precision	8.33	2.95	0.24
DE-EN			
Modified	81.85	87.45	100
Improved	1.62	0.95	0.2
Deteriorated	77.45	84.18	99.65
Precision	1.99	1.09	0.2
EN-ZH			
Modified	97.5	100	100
Improved	26.2	3.5	3.2
Deteriorated	66	93.5	94.4
Precision	29.87	3.5	3.2

4.1.2 Macro and Micro Indicators

In the APE shared task, the submitted systems are analysed via a series of macro and micro indicators (Chatterjee et al. 2020).

Macro indicators (Bojar et al. 2017) measure the number segments which the APE system modifies, improves and deteriorates, using TER with the HPE as a reference. In addition, the precision can also be computed as the 'proportion of improved sentences out of the total amount of modified test items' (Chatterjee et al. 2018, p. 719). A high precision would therefore indicate that an APE model tends to improve the quality of most of the segments it modifies. The macro indicators are reported in Table 3.

Micro indicators correspond to the distribution of edit operations. Here, the edit operations considered correspond to insertions, deletions and substitutions (see Fig. 4).

4.1.3 Human Evaluation

While automatic metrics have several advantages over human evaluation, mostly for reasons of time and cost, it should be acknowledged that they are subject to certain limitations. The most salient one is that they heavily rely on references, which might result in valid translations being penalised. Moreover, they ignore overall grammaticality and synonyms, and assign equal weights to all words (Koehn 2010). Similar concerns can be raised regarding the use of automatic metrics to assess APE.

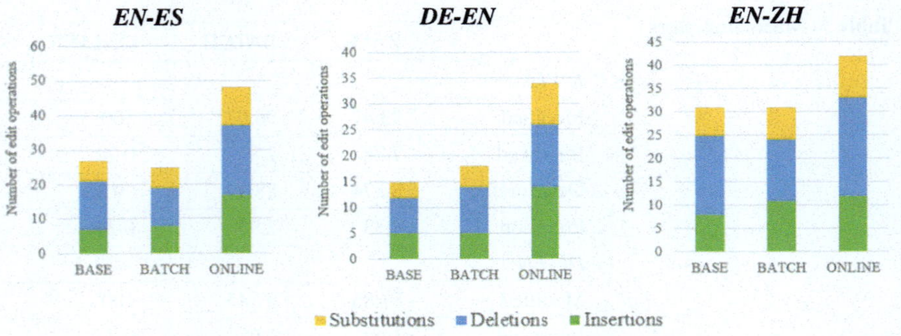

Fig. 4 Micro indicators

Furthermore, human insight is paramount to understand the behaviour of APE models (Pal et al. 2018) and the fact that online APE models described in the literature are analysed only using automatic metrics reinforces the need for human evaluation.

The evaluation procedure is inspired by that of the APE shared task, in which the APE outputs are scored using source-based direct assessment (Chatterjee et al. 2020). Here, 100 segments of each APE model were extracted for human evaluation, resulting in three files per language pair (BASE, BATCH, ONLINE). Evaluators were provided with the source segments and all their corresponding translation options: MTO, APE prediction (BASE, BATCH or ONLINE) and reference HPE. They were asked to assign each candidate translation a score between 0 and 100, considering both adequacy and fluency. To avoid bias, no information regarding the origin of the translations was shared with the participants. In total, 13 participants volunteered (four for EN-ES, three for DE-EN and six for EN-ZH), most of whom have academic or professional experience in translation or are familiar with linguistics. In addition, most of them are native speakers of the relevant target language, and those who are not (only two for DE-EN) are fluent speakers. While the level of these participants was considered high enough to perform the evaluation, this constitutes a limitation, and it is recommended to rely on native speakers in future studies to ensure a thorough analysis of the APE outputs. The scores obtained are reported in Table 4.

4.1.4 Interactive Functionality

The interactive functionality was executed for a few instances in EN-ES to provide a proof of concept. Indeed, for a fair comparison with the other APE models, a translator would need to interactively post-edit at least 1000 sentences, which was not possible for this project. However, a short activity log (one sentence) is included herein for illustration purposes.

Table 4 Human evaluation scores (average scores for each sample of 100 segments)

	BASE	BATCH	ONLINE
EN-ES			
MTO	68	65	60
APE	33	24	1
HPE	78	80	76
DE-EN			
MTO	77	82	71
APE	46	43	2
HPE	94	91	89
EN-ZH			
MTO	50	45	59
APE	19	5	5
HPE	74	73	82

```
Source sentence: That could be Josh someday , speaking before the world 's media ,
beaming out of everyone 's television , one of just over 500 people to ever leave
the orbit of this planet .

Suggested translation: eso podria ser casa algun dia , al hablar cubierto de la
propia comunicacion hasta los medios de comunicacion del mundo , de comparar el
mundo , empece a las la fuerza de comunicacion del mundo , pueden entrar nunca la
fuerza de este planeta .

Are you satisfied with the translation?: (Y for Yes and N for No) N          No

Please provide your translation below:
Eso podría ser Josh algún día, al hablar ante los medios de comunicación del mundo,
ser trasmitido en la televisión de todo el mundo, uno de los poco más de 500
personas a salir de la órbita de este planeta.
```

4.2 Discussion

The results presented in the previous section serve as a basis for answering the research questions formulated for this project. Overall, the results show that all models require significant improvements. Indeed, none of the models could achieve higher scores than the MTO, which is reflected in the outcomes of both automatic metrics and human evaluation.

4.2.1 Batch Models Versus Online Adaptations

No model was able to improve more than 30% of the sentences. The model with the highest precision is the EN-ZH BASE (29.87), while the other two BASE models achieved much lower precision (Table 3). This is a surprising finding: despite being trained with the highest number of samples and an in-domain corpus, the DE-EN BASE appears to have the lowest performance. This phenomenon is also reflected, albeit to a lesser extent, in the BLEU scores (Table 2), with the lowest BLEU for this model as well (though it remains close to the scores of other BASE models). This suggests that the amount of data and the domain do not play crucial roles for training APE models.

BATCH models did not reach a higher precision compared with the BASE models (Table 3). The results of the automatic metrics (Table 2), however, indicate that updating the models in batch mode allowed for yielding the best TER score in EN-ES and the best BLEU score in DE-EN.

Regarding the ONLINE models, these seem to almost yield the lowest results systematically, both in terms of automatic metrics and macro indicators (Tables 2 and 3). The ONLINE model with the highest BLEU is the one for EN-ES, but this score remains significantly lower than that of the baseline MTO for this language pair.

The model which achieved the highest performance overall is the EN-ZH BASE, as it has the highest precision and one of the highest BLEU scores (40, which is slightly below the baseline MTO of 46). This is a considerable achievement, considering that the MTO for this language pair is from an NMT system. Nevertheless, this statement could be toned down, as the EN-ZH language pair was the one with the largest room for improvement. While it has been observed in the shared task that aggressive APE models did not achieve the best performances, here the EN-ZH BASE is also the one with the least conservative behaviour. Another finding which is important to note is that all ONLINE models are systematically aggressive and make the highest number of changes, including all types of edits (Fig. 4). This outcome suggests that a more conservative behaviour might be preferable in the online adaptations.

It should also be highlighted that all models have a high TER, and this score increases systematically in the case of ONLINE models. Overall, the TER moving average plots (Fig. 3) indicate similar behaviours across all language pairs, as all ONLINE models appear to have the worst TER moving average in the long run, whereas the BASE models and the BATCH models seem to be competing.

The outcomes of the human evaluation (Table 4) confirm several of the observations above. In particular, the APE outputs systematically obtained lower scores compared with the MTO baseline and the ONLINE models always yielded the poorest results. This should, however, be treated with caution, as the scores of APE outputs appear to be higher than or equal to those of the MTO baseline on certain occasions (up to 25 times for the DE-EN BATCH). While the EN-ZH BASE was deemed the most efficient overall considering automatic metrics, it does not have the highest score here. Instead, the APE model with the highest score is the DE-EN BASE.

Moreover, the human evaluation allowed for identifying recurrent error patterns in the APE outputs. The most common mistakes consist of inserting information which is absent in the source, deleting content from the original and substituting a word with an incorrect equivalent. Further errors are related to incorrect sentence structures, ambiguities in the source language (e.g. figurative language and MWEs), and these errors being typically exacerbated in the ONLINE models. In addition, the ONLINE models generated more errors, notably repeated words. Another frequent error in outputs from the ONLINE models was to produce translations which were almost unrelated to the source sentence. Nonetheless, the APE models managed to improve the MTO and thus obtained higher scores on a few occasions.

4.2.2 Comparison Results Reported in the Literature

Overall, all systems achieved lower performance compared to the winning systems in the corresponding shared tasks (Bojar et al. 2015, 2017; Chatterjee et al. 2020). One exception, however, is the EN-ZH BASE, which achieved + 2.31 BLEU points compared with the winning system in 2020. As for the online adaptations, the precision, TER and BLEU reported in Chatterjee et al. (2017b) are significantly higher than the scores obtained here. Similarly, the BLEU scores in Negri et al. (2018a) suggest that online APE performs better, both in generic and specialised scenarios. Here, in contrast, the online models all have the lowest performance. The human evaluation scores are also lower compared to those reported in the shared tasks (e.g. 77.2 for EN-ZH in Chatterjee et al. 2020 versus 19 here for the EN-ZH BASE). Several explanations for these gaps in performance can be suggested. In particular, differences in environmental variables compared to the models which served as a basis likely affected the models. Indeed, Chatterjee et al. (2017b) and Negri et al. (2018a) used denser architectures and a significantly higher number of triplets. This observation suggests that the use of synthetic corpora (e.g. eSCAPE) is useful for achieving a higher performance. Similarly, certain environmental variables of the online adaptations (e.g. similarity threshold, number of epochs) most likely also affected the performance. A short experiment in which the EN-ES ONLINE was retrained setting the number of epochs to four instead of three led to an improvement in performance, yielding notably a lower TER (81) and a higher percentage of improved sentences (3.07). These observations call for further research on similar models but with slightly modified variables.

4.2.3 Contributions

The discussion above provides the following answers to the research questions considered for this project.

- *Q1: Does APE yield consistent MT quality improvements across different language pairs?*

  The results are not consistent across language pairs. The overall performance (in terms of precision) can be summarised as follows: EN-ZH > EN-ES > DE-EN.

- *Q2: What is the relation between the original MT output quality and the APE results?*

In this study, none of the APE models improved the MTO.

- *Q3: Which of the analysed APE methods has the highest potential?*

Batch training appears to be the most appropriate for APE models, with the best-performing system being a BASE (EN-ZH).

- *Q4: Is online adaptation an efficient strategy to improve the performance of APE?*

It appears that online adaptation does not provide efficient improvements. This finding is nuanced, however, as the performance is likely to have suffered from environmental variables and data size.

- *Q5: How do APE models trained on non-synthetic data perform?*

Given that the systems developed for this study yield a poorer performance compared to similar systems trained on synthetic corpora, it is recommended to utilise such data when training APE systems.

It is also important to make further comments on the interactive functionality. A brief demonstration was provided using the command line to interact with the user, which is not a realistic scenario. Therefore, although doing so was judged to be beyond the scope of this project, designing an interface for collecting HPE on the fly and dynamically using this data for online training are strongly recommended approaches for future research (Escribe and Mitkov 2021). Such an interface would allow to find whether interactive PE models can adapt to the style of the translator and the document being translated (ibid.). Furthermore, this type of tool would be likely to enhance the experience of translators by reducing the PE effort, which could be studied via various measures (e.g. questionnaires, keylogging, eye tracking).

Finally, the present project makes several contributions to the field of APE, as it:

i. Studies the performance of online adaptations of neural APE models on outputs from both STM and NMT systems
ii. Analyses the performance of neural APE models updated in both batch and online modes for three distinct language pairs (EN-ES, DE-EN, EN-ZH) and three different domains (news, pharmacological, Wikipedia) without using any synthetic data
iii. Provides a human evaluation of the outputs obtained in all training modes, and in particular online adaptations (which tend to be evaluated using automatic metrics only), thus providing further insight into the APE outputs
iv. Introduces an interactive functionality which lets the translator dynamically enter post-edits

It is evident from the scores obtained that the performance of the systems herein remained rather low, which confirms that PE is particularly difficult to automate.

5 Conclusion

This study examined the performance of APE models in both batch and online training modes, considering various language pairs and domains without using any synthetic corpus. What emerges from the results is that none of the APE models developed could outperform the baseline, and that online adaptations systematically yielded a lower performance. This outcome could be partly attributed to certain environmental variables and the limited data size. Overall, denser NNs, a higher number of epochs and more data (possibly synthetic corpora) are recommended for follow-up projects.

Another important finding is that a generic domain and NMT outputs, which first appear as additional difficulties for APE models, do not impede improvements, as the EN-ZH BASE reached a higher performance than the winning system of the 2020 APE shared task.

Furthermore, the human evaluation provided observations serving to identify recurrent error patterns in the APE outputs which are exacerbated in the online adaptations, including notably incorrect deletions, insertions and substitutions, and errors related to sentence structure and figurative language.

In addition, the present study introduced an interactive functionality in which translators can dynamically enter post-edits, which are in turn fed back into the system to update its parameters on the fly.

Overall, the limited gains achieved by APE models confirm the difficulty of automating PE. To date, no system has been able to fully automate the PE process, and APE outputs require further corrections by translators to reach a satisfactory quality. Based on this observation, conceptualising PE models as interactive systems rather than seeking to fully automate the PE process emerges as a recommendable option.

Corpora are the cornerstone of latest innovations in translation technology. The findings discussed herein therefore constitute valuable contributions in this field and in other related study domains. Assuming that developments in APE research yield more satisfactory results, this technology could become an integral part of translation workflows. Hence, a rigorous methodology for a closer examination of APE corpora and clear guidelines for compiling training data would be highly beneficial for APE research. In-depth linguistic assessment of APE outputs (including batch, online and interactive models) are also recommended to achieve an extensive understanding of APE performance.

Acknowledgements We would like to express our sincere gratitude to the volunteers who kindly accepted to work on the evaluation tasks, in particular:

• *the English–Spanish team*: Lucía Bellés-Calvera (Universitat Jaume I), Rocío Caro Quintana (University of Wolverhampton), Ana Isabel Cespedosa Vázquez (Universidad de Córdoba) and Ana Isabel Martínez Hernández (Universitat Jaume I).

• *the German-English team*: Anne Eschenbrücher (University of Wolverhampton), Lydia Körber (University of Potsdam, Free University of Berlin) and Alistair Plum (University of Wolverhampton).

• *the English-Chinese team*: Chien-yu Chen (University of Barcelona), Jacinda Chen (Hong Kong Polytechnic University), Meng Chunyu (Hong Kong Baptist University), Zhujun Zhang (Soochow University), Hellen Zheng (Anastacio Overseas Inc.) and Ruiqi Zhou (Hong Kong Baptist University).

References

Alabau, Vicent, Ragnar Bonk, Christian Buck, Michael Carl, Francisco Casacuberta, Mercedes García-Martínez, Jesús González, Philipp Koehn, Luis Leiva, Bartolomé Mesa-Lao, Daniel Ortiz, Herve Saint-Amand, Germán Sanchis, and Chara Tsoukala. 2013. CASMACAT: An open source workbench for advanced computer aided translation. *The Prague Bulletin of Mathematical Linguistics* 100 (1): 101–112.

Allen, Jeffrey, and Christopher Hogan. 2000. Toward the development of a post editing module for raw machine translation output: A controlled language perspective. In *Proceedings of the 3rd International Controlled Language Applications Workshop (CLAW-00)*, 62–71. Seattle, Washington, USA.

Alves, Fabio, Arlene Koglin, Bartolomé Mesa-Lao, Mercedes García-Martínez, Norma B. de Lima Fonseca, Arthur de Melo Sá, José Luiz Gonçalves, Karina Sarto Szpak, Kyoko Sekino, and Marceli Aquino. 2016a. Analysing the impact of interactive machine translation on post-editing effort. In *New Directions in Empirical Translation Process Research*, ed. Michael Carl, Srinivas Bangalore, and Moritz Schaeffer, 77–94. Cham: Springer.

Alves, Fabio, Karina Sarto Szpak, José Luiz Gonçalves, Kyoko Sekino, Marceli Aquino, Rodrigo Araújo e Castro, Arlene Koglin, Norma B. de Lima Fonseca, and Bartolomé Mesa-Lao. 2016b. Investigating cognitive effort in post-editing: A relevance-theoretical approach. In *Eyetracking and Applied Linguistics*, ed. Silvia Hansen-Schirra, and Sambor Grucza, 109–142. Berlin: Language Science Press.

Bahdanau, Dzmitry, KyungHyun Cho, and Yoshua Bengio. 2014. Neural machine translation by jointly learning to align and translate. *Preprint*.

Bar-Hillel, Yehoshua. 1960. The present status of automatic translation of languages. *Advances in Computers*. 1: 91–163.

Bojar, Ondřej, Rajen Chatterjee, Christian Federmann, Barry Haddow, Matthias Huck, Chris Hokamp, Philipp Koehn, Varvara Logacheva, Christof Monz, Matteo Negri, Matt Post, Carolina Scarton, Lucia Specia, and Marco Turchi. 2015. Findings of the 2015 workshop on statistical machine translation. In *Proceedings of the 10th Workshop on Statistical Machine Translation*, 1–46. Lisboa, Portugal.

Bojar, Ondřej, Rajen Chatterjee, Christian Federmann, Yvette Graham, Barry Haddow, Matthias Huck, Antonio Jimeno Yepes, Philipp Koehn, Varvara Logacheva, Christof Monz, Matteo Negri, Aurélie Névéol, Mariana Neves, Martin Popel, Matt Post, Raphael Rubino, Carolina Scarton, Lucia Specia, Marco Turchi, Karin Verspoor, and Marcos Zampieri. 2016. Findings of the 2016 conference on machine translation (WMT16). In *Proceedings of the 1st Conference on Machine Translation, Volume 2: Shared Task Papers*, 131–198. Berlin, Germany.

Bojar, Ondřej, Rajen Chatterjee, Christian Federmann, Yvette Graham, Barry Haddow, Huang Shujian, Matthias Huck, Philipp Koehn, Qun Liu, Varvara Logacheva, Christof Monz, Matteo Negri, Matt Post, Raphael Rubino, Lucia Specia, and Marco Turchi. 2017. Findings of the 2017 conference on machine translation (WMT17). In *Proceedings of the Conference on Machine Translation, Volume 2: Shared Task Papers*, 169–214. Copenhagen, Denmark.

Castaño, Asuncion, and Francisco Casacuberta. 1997. A connectionist approach to machine translation. In *Proceedings of the 5th European Conference on Speech Communication and Technology*, 221–229. Rhodes, Greece.

Castilho, Sheila. 2020. Document-level machine translation evaluation project: Methodology, effort and inter-annotator agreement. In *Proceedings of the 22nd Annual Conference of the European Association for Machine Translation*, 475–476. Lisbon, Portugal.

Castilho, Sheila, Joss Moorkens, Federico Gaspari, Iacer Calixto, John Tinsley, and Andy Way. 2017. Is neural machine translation the new state of the art? *The Prague Bulletin of Mathematical Linguistics* 108 (1): 109–120.

Chatterjee, Rajen, Marion Weller, Matteo Negri, and Marco Turchi. 2015. Exploring the planet of the APEs: A comparative study of state-of-the-art methods for MT automatic post-editing. In *Proceedings of the 53rd Annual Meeting of the Association for Computational Linguistics and the 7th International Joint Conference on Natural Language Processing, Volume 2: Short Papers*, 156–161. Beijing, China.

Chatterjee, Rajen, Amin Farajian, Matteo Negri, Marco Turchi, Ankit Srivastava, and Santanu Pal. 2017a. Multi-source neural automatic post-editing: FBK's participation in the WMT 2017 APE shared task. In *Proceedings of the Conference on Machine Translation (WMT), Volume 2: Shared Task Papers*, 630–638. Copenhagen, Denmark.

Chatterjee, Rajen, Gebremedhen Gebremelak, Matteo Negri, and Marco Turchi. 2017b. Online automatic post-editing for MT in a multi-domain translation environment. In *Proceedings of the 15th Conference of the European Chapter of the Association for Computational Linguistics: Volume 1, Long Papers*, 525–535. Valencia, Spain.

Chatterjee, Rajen, Matteo Negri, Raphael Rubino, and Marco Turchi. 2018. Findings of the WMT 2018 shared task on automatic post-editing. In *Proceedings of the 3rd Conference on Machine Translation (WMT), Volume 2: Shared Task Papers*, 723–738. Brussels, Belgium.

Chatterjee, Rajen, Christian Federmann, Matteo Negri, and Marco Turchi. 2019. Findings of the WMT 2019 shared task on automatic post-editing. In *Proceedings of the 4th Conference on Machine Translation (WMT), Volume 3: Shared Task Papers (Day 2)*, 13–30. Florence, Italy.

Chatterjee, Rajen, Markus Freitag, Matteo Negri, and Marco Turchi. 2020. Findings of the WMT 2020 shared task on automatic post-editing. In *Proceedings of the 5th Conference on Machine Translation (WMT)*, 646–659. Online.

Cho, Kyunghyun, Bart van Merriënboer, Caglar Gulcehre, Dzmitry Bahdanau, Fethi Bougares, Holger Schwenk, and Yoshua Bengio. 2014. Learning phrase representations using RNN encoder–decoder for statistical machine translation. In *Proceedings of the 2014 Conference on Empirical Methods in Natural Language Processing (EMNLP)*, 1724–1734. Doha, Qatar.

do Carmo, Félix, Dimitar Shterionov, Joss Moorkens, Joachim Wagner, Murhaf Hossari, Eric Paquin, Dag Schmidtke, Declan Groves, and Andy Way. 2020. A review of the state-of-the-art in automatic post-editing. *Machine Translation* 35: 101–143.

Domingo, Miguel, Mercedes García-Martínez, Álvaro Peris, Alexandre Helle, Amando Estela, Laurent Bié, Francisco Casacuberta, and Manuel Herranz. 2019. Incremental adaptation of NMT for professional post-editors: A user study. In *Proceedings of Machine Translation Summit XVII, Volume 2: Translator, Project and User Tracks*, 219–227. Dublin, Ireland.

Domingo, Miguel, Mercedes García-Martínez, Álvaro Peris, Alexandre Helle, Amando Estela, Laurent Bié, Francisco Casacuberta, and Manuel Herranz. 2020. A user study of the incremental learning in NMT. In *Proceedings of the 22nd Annual Conference of the European Association for Machine Translation*, 319–328 (Online).

Escribe, Marie, and Ruslan Mitkov. 2021. Interactive models for post-editing. In *Proceedings of TRITON (Translation and Interpreting Technology Online)*, 167–173 (Online).

Esteban, José, José Lorenzo, Antonio S. Valderrábanos, and Guy Lapalme. 2004. TransType2—An innovative computer-assisted translation system. In *Proceedings of the ACL Interactive Poster and Demonstration Sessions*, 94–97. Barcelona, Spain.

Farajian, M. Amin, Marco M. Turchi, Matteo Negri, and Marcello Federico. 2017. Multi-domain neural machine translation through unsupervised adaptation. In *Proceedings of the Conference on Machine Translation (WMT), Volume 1: Research Papers*, 127–137. Copenhagen, Denmark.

Foster, George, Pierre Isabelle, and Pierre Plamondon. 1997. Target-text mediated interactive machine translation. *Machine Translation* 12 (1): 175–194.

Foster, George, Philippe Langlais, and Guy Lapalme. 2002. User-friendly text prediction for translators. In *Proceedings of the 2002 Conference on Empirical Methods in Natural Language Processing (EMNLP 2002)*, 148–155. Philadelphia, Pennsylvania, USA.

Green, S., Jeffrey Heer, and Christopher D. Manning. 2015. Natural language translation at the intersection of AI and HCI. *Communications of the ACM* 58 (9): 46–53.

Guerreiro, Nuno M., Elena Voita, and F.T. André Martins. 2022. Looking for a needle in a haystack: A comprehensive study of hallucinations in neural machine translation. *Preprint*.

Hassan, Hany Anthony Aue, Chang Chen, Vishal Chowdhary, Jonathan Clark, Christian Federmann, Xuedong Huang, Marcin Junczys-Dowmunt, William Lewis, Mu Li, Shujie Liu, Tie-Yan Liu, Renqian Luo, Arul Menezes, Tao Qin, Frank Seide, Xu Tan, Fei Tian, Lijun Wu, Shuangzhi Wu, Yingce Xia, Dongdong Zhang, Zhirui Zhang, Ming Zhou. 2018. Achieving human parity on automatic Chinese to English news translation. *Microsoft AI & Research*.

Hinton, Geoffrey E., Nitish Srivastava, Alex Krizhevsky, Ilya Sutskever, and Ruslan Salakhutdinov. 2012. Improving neural networks by preventing co-adaptation of feature detectors. *Preprint*.

Hutchins, John. 2005. *The First Public Demonstration of Machine Translation: The Georgetown-IBM System, 7th January 1954*.

Junczys-Dowmunt, Marcin. 2018. Are we experiencing the golden age of automatic post-editing? In *Proceedings of the AMTA 2018 Workshop on Translation Quality Estimation and Automatic Post-Editing*, 144–206. Boston, Massachusetts, USA.

Karimova, Sariya, Patrick Simianer, and Stefan Riezler. 2018. A user-study on online adaptation of neural machine translation to human post-edits. *Machine Translation* 32: 309–324.

Kingma, Diederik P., and Jimmy Lei Ba. 2015. Adam: A method for stochastic optimisation. *Preprint*.

Knight, Kevin, and Ishwar Chander. 1994. Automated post-editing of documents. In *Proceedings of the 12th National Conference on Artificial Intelligence (AAAI)*, 779–784. Seattle, Washington, USA.

Knowles, Rebecca, and Philipp Koehn. 2016. Neural interactive translation prediction. In *Proceedings of the Association for Machine Translation in the Americas*, 107–120. Austin, Texas, USA.

Koehn, Philipp. 2010. *Statistical Machine Translation*. Cambridge: Cambridge University Press.

Lagarda, Antonio L., Daniel Ortiz-Martínez, Vincent Alabau, and Francisco Casacuberta. 2015. Translating without in-domain corpus: Machine translation post-editing with online learning techniques. *Computer Speech and Language*. 32 (1): 109–134.

Lagoudaki, Elina. 2008. The value of machine translation for the professional translator. In *Proceedings of the 8th Conference of the Association for Machine Translation in the Americas (AMTA)*, 262–269. Waikiki, Hawaii, USA.

Licklider, Joseph C.R. 1960. Man-computer symbiosis. *IRE Transactions on Human Factors in Electronics* 1: 4–11.

Mitkov, Ruslan. 2021. Translation memory. In *The Routledge Handbook of Translation and Memory*, ed. Sue-Ann Deane-Cox, and Anneleen Spiessens. Basingstoke: Routledge.

Ñeco, Ramón P., and Mikel L. Forcada. 1997. Asynchronous translations with recurrent neural nets. In: *Proceedings of the IEEE International Conference on Neural Networks*, vol. 4, 2535–2540. Houston, Texas, USA.

Negri, Matteo, Marco Turchi, Nicola Bertoldi, and Marcello Federico. 2018a. Online neural automatic post-editing for neural machine translation. In *Proceedings of the 5th Italian Conference on Computational Linguistics (CLIC-IT 2018)*, 288–293. Torino, Italy.

Negri, Matteo, Marco Turchi, Rajen Chatterjee, and Nicola Bertoldi. 2018b. eSCAPE: A large-scale synthetic corpus for automatic post-editing. In *Proceedings of the 11th International Conference on Language Resources and Evaluation (LREC 2018)*, 24–30. Miyazaki, Japan.

Nishida, Fujio, Shinobu Takamatsu, Tadaaki Tani, and Tsunehisa Doi. 1988. Feedback of correcting information in post-editing to a machine translation system. In *Proceedings of the 12th International Conference on Computational Linguistics*, vol. 2, 476–481. Budapest, Hungary.

Ortiz-Martínez, Daniel, and Francisco Casacuberta. 2014. The new Thot toolkit for fully-automatic and interactive statistical machine translation. In *Proceedings of the Demonstrations at the 14th Conference of the European Chapter of the Association for Computational Linguistics*, 45–48. Gothenburg, Sweden.

Ortiz-Martínez, Daniel, Ismael García-Varea, and Francisco Casacuberta. 2010. Online learning for interactive statistical machine translation. In *Proceedings of Human Language Technologies: The 2010 Annual Conference of the North American Chapter of the ACL*, 546–554. Los Angeles, California, USA.

Pal, Santanu, Sudip Kumar Naskar, Mihaela Vela, Qun Liu, and Josef van Genabith. 2017. Neural automatic post-editing using prior alignment and reranking. In *Proceedings of the 15th Conference of the European Chapter of the Association for Computational Linguistics: Volume 2, Short Papers*, 349–355. Valencia, Spain.

Pal, Santanu, Nico Herbig, Antonio Krüger, and Josef van Genabith. 2018. A transformer-based multi-source automatic post-editing system. In *Proceedings of the 3rd Conference on Machine Translation: Shared Task Papers*, 827–835. Brussels, Belgium.

Papineni, Kishore, Salim Roukos, Todd Ward, and Zhu Wei-Jing. 2002. BLEU: A method for automatic evaluation of machine translation. In *Proceedings of the 40th Annual Meeting of the Association for Computational Linguistics*, 311–318. Philadelphia, Pennsylvania, USA.

Peris, Álvaro, and Francisco Casacuberta. 2019. Online learning for effort reduction in interactive neural machine translation. *Computer Speech and Language* 58 (1): 98–126.

Phaholphinyo, Sitthaa, Teerapong Modhiran, Nattapol Kritsuthikul, and Thepchai Supnithi. 2005. A practical of memory-based approach for improving accuracy of MT. In *Proceedings of the MT Summit X*, 41–46. Phuket, Thailand.

Poibeau, Thierry. 2022. On "human parity" and "super human performance" in machine translation evaluation. In *Proceedings of the Language Resource and Evaluation Conference*, 6018–6023. Marseille, France.

Salton, Gerard, Chung-Shu Yang, and T. Yu. Clement. 1975. A theory of term importance in automatic text analysis. *Journal of the American Society for Information Science* 26: 33–44.

Santy, Sebastin, Sandipan Dandapat, Monojit Choudhury, and Kalika Bali. 2019. INMT: Interactive neural machine translation prediction. In *Proceedings of the 2019 Conference on Empirical Methods in Natural Language Processing and the 9th International Joint Conference on Natural Language Processing (EMNLP-IJCNLP): System Demonstrations*, 103–108. Hong Kong, China.

Shterionov, Dimitar, Félix do Carmo, Joss Moorkens, Murhaf Hossari, Joachim Wagner, Eric Paquin, Dag Schmidtke, Declan Groves, and Andy Way. 2020. A roadmap to neural automatic post-editing: An empirical approach. *Machine Translation* 34 (2): 67–96.

Simard, Michel, and George Foster. 2013. PEPr: Post-edit propagation using phrase-based statistical machine translation. In *Proceedings of the XIV Machine Translation Summit*, 191–198. Nice, France.

Simard, Michel, Cyril Goutte, and Pierre Isabelle. 2007. Statistical phrase-based post-editing. In *Human Language Technologies 2007: The Conference of the North American Chapter of the Association for Computational Linguistics; Proceedings of the Main Conference (NAACL HLT)*, 508–515. Rochester, New York, USA.

Snover, Matthew, Bonnie Dorr, Richard Schwartz, Linnea Micciulla, and John Makhoul. 2006. A study of translation edit rate with targeted human annotation. In *Proceedings of the 7th Conference of the Association for Machine Translation of the Americas*, 223–231. Cambridge, Massachusetts, USA.

Su, Keh-Yih, Jing-Shin Chang, and Yu-Ling Una Hsu. 1995. A corpus-based statistics-oriented two-way design for parameterised MT systems: Rationale, architecture and training issues. In *Proceedings of the 6th International Conference on Theoretical and Methodological Issues in Machine Translation*, 334–353. Leuven, Belgium.

Toral, Antonio, Sheila Castilho, Ke Hu, and Andy Way. 2018. Attaining the unattainable? Reassessing claims of human parity in neural machine translation. In *Proceedings of the 3rd Conference on Machine Translation: Research Papers*, 113–123. Brussels, Belgium.

Toselli, Alejandro Héctor, Enrique Vidal, and Francisco Casacuberta. 2011. *Multimodal Interactive Pattern Recognition and Applications*. London: Springer Science & Business Media.

Underwood, Nancy, Bartolomé Mesa-Lao, Mercedes García Martínez, Michael Car, Vicent Alabau, Jesús González-Rubio, Luis A. Leiva, Germán Sanchis-Trilles, Daniel Ortíz-Martínez, Francisco Casacuberta. 2014. Evaluating the effects of interactivity in a post-editing workbench. In *Proceeding of the 9th International Conference on Language Resources and Evaluation (LREC)*, 553–559. Reykjavik, Iceland.

Vaswani, Ashish, Noam Shazeer, Niki Parmar, Jakob Uszkoreit, Llion Jones, Aidan N. Gomez, Łukasz Kaiser, Illia Polosukhin. 2017. Attention is all you need. In *Proceedings of the 31st Annual Conference on Advances in Neural Information Processing Systems (NIPS)*, 5998–6008. Long Beach, California, USA.

Zaninello, Andrea, and Alexandra Birch. 2020. Multiword expression aware neural machine translation. In *Proceedings of the 12th Language Resources and Evaluation Conference*, 3816–3825. Marseille, France.

Zhechev, Ventsislav. 2012. Machine translation infrastructure and post-editing performance at Autodesk. In *Proceedings of the Workshop on Post-editing Technology and Practice*. San Diego, California, USA.

Marie Escribe is a Ph.D. student in Translation Technology at the Polytechnic University of Valencia and a linguistic engineer at LanguageWire. She has several years of experience as a translator, and holds an M.A. in Translation from London Metropolitan University and an M.A. in Computational Linguistics from the University of Wolverhampton. Her research interests revolve around translation technologies and include in particular post-editing, computer-assisted translation tools, translation memory systems and translation quality evaluation.

Ruslan Mitkov has been working in Natural Language Processing (NLP), Computational Linguistics, Corpus Linguistics, Machine Translation, Translation Technology and related areas since the early 1980s. Whereas Prof Mitkov is best known for his seminal contributions to the areas of anaphora resolution and automatic generation of multiple-choice tests, his extensively cited research (more than 260 publications including 17 books, 40 journal articles and 40 book chapters) also covers topics such as machine translation, translation memory and translation technology in general, bilingual term extraction, automatic identification of cognates and false friends, natural language generation, automatic summarisation, computer-aided language processing, centring, evaluation, corpus annotation, NLP-driven corpus-based study of translation universals, text simplification, NLP for people with language disabilities and computational phraseology. Current topics of research interest include the employment of deep learning techniques in translation and interpreting technology as well as conceptual difficulty for text processing and translation.

Integrating Trados-Qualitivity Data to the CRITT TPR-DB: Measuring Post-editing Process Data in an Ecologically Valid Setting

Longhui Zou, Michael Carl, and Devin Gilbert

1 Introduction

Higher quality of machine translation (MT) output and greater availability of computer-aided translation (CAT) tools has led to the prevalence of post-editing of MT (PEMT) in today's language industry (Sun 2019). PEMT often involves a workflow that includes machine translating the source text (ST) and subsequently having a human translator post-edit the MT output in order to create a finished translation product. Using MT has generally been found to improve translation productivity compared to translation from scratch. The recent rise and dominance of neural machine translation (NMT) has been shown to produce translations with greater fluency and accuracy compared to other MT paradigms, thus facilitating the translation process, and increasing translator efficiency and productivity (Bentivogli et al. 2016; Toral and Sánchez-Cartagena 2017; Toral et al. 2018).

Furthermore, CAT tools such as Trados Studio or memoQ help translators to produce more translated words per unit of time whilst facilitating quality assurance and quality control. Utilising CAT tools has become the norm in the world of professional translators since they can boost translation productivity by providing not only translation memories (TM), which help translators reuse legacy translations, but also allow for consulting terminology databases (TB) and concordance tools, which help maintain consistency throughout the entire translation process and across different translators within and between translation projects (Vieira et al. 2021).

L. Zou (✉) · M. Carl
Kent State University, Kent, USA
e-mail: lzou4@kent.edu

M. Carl
e-mail: mcarl6@kent.edu

D. Gilbert
Utah Valley University, Orem, USA
e-mail: devinG@uvu.edu

J. Pan and S. Laviosa (eds.), *Corpora and Translation Education*, New Frontiers in Translation Studies, https://doi.org/10.1007/978-981-99-6589-2_4

Language service providers (LSPs) have also turned to CAT tools to collect behavioural data of translators and post-editors. Keyloggers are programmes, running in the background of the computer to record user activity data (UAD) from a keyboard or other textual input devices (Leijten and Van Waes 2013). Keyloggers record keystrokes, mouse clicks, scrolls, and pauses in between key presses along with their corresponding timestamps. These resources are often used by LSPs to measure translators' production speed to compute the edit distance, track the productivity, and progress of translation projects, profile translators' performance, determine pricing, and to calculate rates and payment for translators' work.

For instance, the translation agency Translated.net uses a free browser-based, open-source CAT tool, MateCat, for their translation projects with the purpose of supporting PEMT. MateCat can gather timing information, generate suggestions from MT or the TM, and record the actual editing operations and keystrokes for each segment. Such information can be accessed together with detailed statistics such as percentage of fuzzy match scores, PEMT effort, and translation speed through a summary page called Edit Log (Federico et al. 2014).

Apart from browser-based CAT tools (e.g., MateCat, Unbabel, Lilt, etc.) that make it possible to track and analyse the translation process (Elming and Balling 2014), also Trados Studio has a plugin—called Qualitivity—that collects keystroke activities of the translator.[1] The functionalities of Qualitivity make it possible to capture every single keystroke, and the changes these keystrokes produce in a target segment, during translating, reviewing, or post-editing. The Qualitivity plugin also records the time spent on these translation activities on the segment level and supports four Quality Metric standards (e.g., TAUS DQF, MQM Core, SAE J2450, LISA QA Model) that allow for customised statistical information; therefore, this plugin can aid data management when it comes to PEMT productivity and quality.

PEMT is increasingly in the focus of different fields of translation studies, including in translation process research (TPR). Most scholars focus on operationalising and measuring the effort involved in post-editing (Bentivogli et al. 2018; Carl et al. 2019; Läubli et al. 2020; Toral et al. 2018; Vieira et al. 2019). In terms of effort measurement, Krings's framework has been widely applied by TPR scholars (Krings 2001). It proposes three types of post-editing effort: temporal effort the speed or productivity rate of post-editing, for instance, words per second (Moorkens 2018); technical effort (the number of keystrokes or actual edits conducted by the post-editor, for instance, Translation Edit Rate (Snover et al. 2006); and cognitive effort (the underlying mental effort that ties in with both temporal and technical effort, for instance, Pause to Word Ratio (Lacruz et al. 2014).

Whilst temporal and technical effort has mostly been measured by keylogging, alternative research methods such as Think-Aloud Protocol (TAP), eye-tracking, Functional Magnetic Resonance Imaging (fMRI), Electroencephalography (EEG) have also been used either alone or combined with other methods to investigate cognitive effort in TPR (Jakobsen 1999; O'Brien 2006; Chang 2009; Dragsted 2010; Hvelplund 2011; Vieira 2016; Hansen-Schirra 2017; Walker and Federici 2018). In

[1] See https://appstore.rws.com/Plugin/16.

the last decade, however, eye-tracking has become an essential method for research on cognitive effort in translation, particularly in PEMT (Lacruz 2017).

The common hypothesis underlying TPR employing eye-tracking is the Eye-Mind assumption, which posits that an object which is fixated is almost simultaneously processed in one's brain (Just and Carpenter 1980). This suggests a straightforward relationship between the duration and count of fixations and cognitive effort in the translation process. Eye-trackers capture the fixations of the participant as "eye movements which stabilise the retina over a stationary object of interest" (Duchowski 2003). It can provide large amount of data resources such as the number and sequence of fixations, the areas of interest (AOIs) with most frequent fixations, the video recordings with the eye gaze, the X, Y coordinates of the gaze positions on the screen, and left and right pupil sizes, etc. (O'Brien 2009).

A large amount of TPR has been conducted with Translog-II. Translog-II is a research tool that makes it possible to record keystrokes and gaze data during translation sessions (Carl 2012). The collected data can then be uploaded to the CRITT Translation Process Research Database (CRITT TPR-DB) which provides numerous tools for data analysis and data visualisation (Carl et al. 2016). However, Translog-II is a research tool that does not offer a translation environment that professional translators are used to. Professional translators work in more complex computer-aided translation (CAT) tools. Experimental research with Translog-II may therefore bring about generalisations from the laboratory setting that should be validated with non-experimental, real-world translation behaviour. This discrepancy between the experimental environment and translators' real-world working environments can be considered an experimental problem that violates ecological validity. Ecological validity refers to how similar an experimental setting is to the actual real-world conditions of the phenomenon that is being studied (Mellinger and Hanson 2022).

To simulate the real-world translation work environment and to increase the ecological validity of TPR, we present a new interface (i.e., Trados-to-Translog-II) that integrates keystroke and eye-tracking data collected within Trados Studio with the data format required by the CRITT TPR-DB (Zou and Carl 2022; Yamada et al. 2022). The Qualitivity plugin has already been available for several years, but the Trados-to-Translog-II interface makes it possible, for the first time, to use data collected with the Qualitivity plugin in the CRITT TPR-DB.

Sections 2–4 are essentially a sort of user manual for other researchers who would like to conduct TPR research in Trados Studio and later integrate their data with the CRITT TPR-DB. In Sect. 2, we first discuss how to record keystrokes in Trados Studio using the Qualitivity plugin. In Sect. 3, we demonstrate how the keystroke log files that were exported from the Qualitivity plugin, can be converted to a format so that the data can be uploaded to the CRITT TPR-DB. In Sect. 4, we show how an eye-tracker can be used for the same Trados Studio text production sessions and its data synchronised with the Qualitivity log file.

In Sect. 5, we present a remote post-editing study with Trados Studio. It elicits data collection principles in a remote setting and outlines our experience that are crucial to successfully gathering CRITT TPR-DB-compatible data in Trados Studio. Subsequently, we present a small study framework in Sect. 6 that triangulates Trados keylogging data and Tobii eye-tracking data within the CRITT TPR-DB. In Sect. 7, we discuss pedagogical applications for word-aligned parallel corpora produced by professional translators and housed in the CRITT TPR-DB and conclude the article in Sect. 8.

2 Recording Keystrokes in Trados Studio Using the Qualitivity Plugin

The new Trados-to-Translog-II interface allows to collect key-logging data from Trados in order to analyse within the CRITT TPR-DB. A Trados-to-Translog-II post-editing experiment involves the following steps:

1. Download the Qualitivity plugin for Trados Studio, version 2019, and install the plugin. Integration of Qualitivity keystroke logs has only been tested for Trados Studio 2019, so this version constraint is imperative.
2. Open Trados Studio. Successful installation results in a new "Qualitivity" navigation button in the bottom left corner of Trados Studio interface as shown in Fig. 1.
3. Click the "Projects" navigation button and set up a new post-editing project for one participant. This would be the same process as a routinised translation session working in Trados Studio, as shown in Fig. 2. Once the project has been created, a Qualitivity dialog box will automatically pop up asking if you want to create a Qualitivity project to accompany the Trados Studio project that was just created. Whether or not you create this Qualitivity project beforehand, Qualitivity will automatically log the keystrokes during a text production session within the Trados Studio editor. Qualitivity only records keystrokes that cause a character

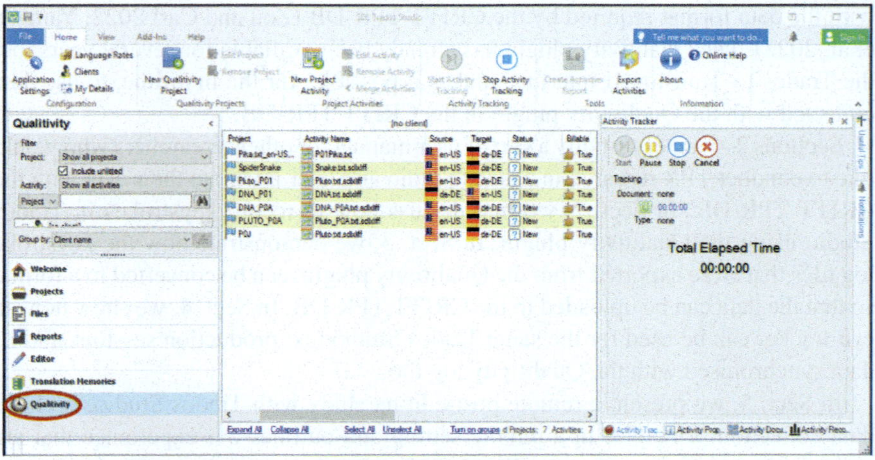

Fig. 1 Qualitivity navigation button in Trados Studio

in an active segment to either be inserted or deleted. Additionally, it records text that is automatically populated from the TM or from MT APIs; however, it only records automatically populated text for the *active* segment.

a. A segment is considered "active" in Trados Studio if the cursor is within it. It does not matter how the cursor gets there (the user clicks inside the segment with the mouse; they confirm the previous segment, which in most cases, automatically activates the next segment; they perform a "find" operation for text that the segment contains, etc.).
b. Keystroke combinations that do not result in text being inserted or deleted will not be recorded.
c. Text inserted by pasting using the mouse or a keyboard shortcut will be recorded.
d. Text inserted from the TM or an MT suggestion will only be recorded for active segments, which means that text inserted during pre-translation or look-ahead operations will not show up in keystroke logs produced by Qualitivity.

Because Qualitivity only records inserted text for active segments, special care must be taken to make sure Qualitivity keystroke logs do not lack data in the event users decide not to make any edits to a particular segment. For experimental text production sessions in Trados Studio, one should not pre-translate files because there is a high risk that users will neglect to place their cursor within the segments they do not wish to edit. This means they will not "activate" segments they are not editing, and any segments that were never made *active* during a text production session will be missing from the resulting Qualitivity keystroke log. If a segment is missing from the

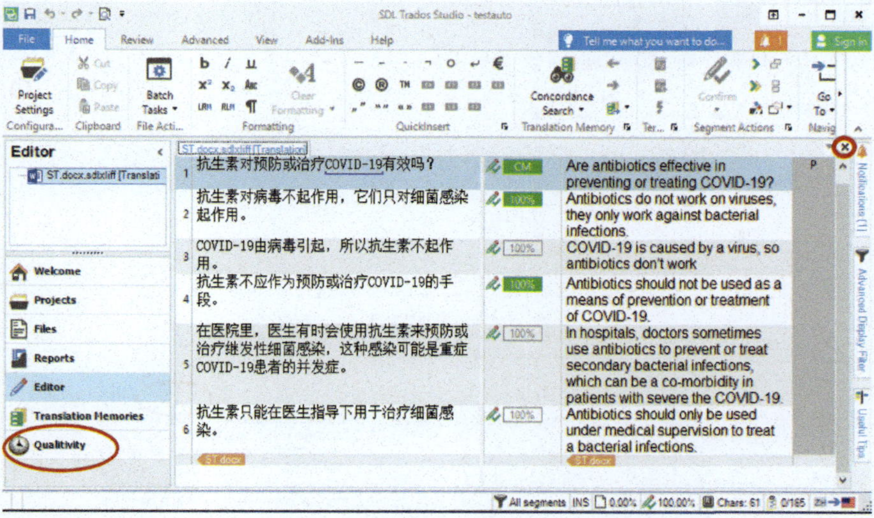

Fig. 2 Post-editing interface in Trados Studio

Qualitivity keystroke log, then it will also be missing from the converted Translog file. This data loss is extremely hard to fix manually and therefore must be prevented. **As long as files are not pre-translated, and Trados Studio settings are such that TM matches or MT suggestions only populate once a user has *activated* a segment, chances of users simply not activating a segment are reliably low.** If one wishes to integrate their data with the CRITT TPR-DB, these settings are imperative for a successful experiment using Trados Studio (these settings are discussed in detail in Sect. 5).

4. After the participant completes the entire post-editing experiment, exit the Trados Studio editor by hitting the "X" button in the upper-right corner of the editor as shown by the red circle on the right side in Fig. 2 (not the "X" button in the upper-right corner of the Trados Studio programme window), and save the Qualitivity activity by hitting "OK" in the pop-up dialog box, as shown in Fig. 3.
5. Export the keylogging data collected by Qualitivity plugin to an XML format as shown in Fig. 4. Make sure to tick the box "Include keystroke data" and select "Export to XML format". From the XML file output, we can see all the post-editor's changes to the MT with timestamps recording the exact production time and position of each edit. For example, the source text of one segment in this sample experiment in Chinese is: "在医院里, 医生有时会使用抗生素来预防或治疗继发性细菌感染, 这种感染可能是重症。", and the target text (MT) in English is: "In hospitals, doctors sometimes use antibiotics to prevent or treat secondary bacterial infections, which can be a co-morbidity in patients with severeCOVID-19". The post-editor in the sample experiment added a space between "severe" and "COVID-19" and added the word "would" between "sometimes" and "use".

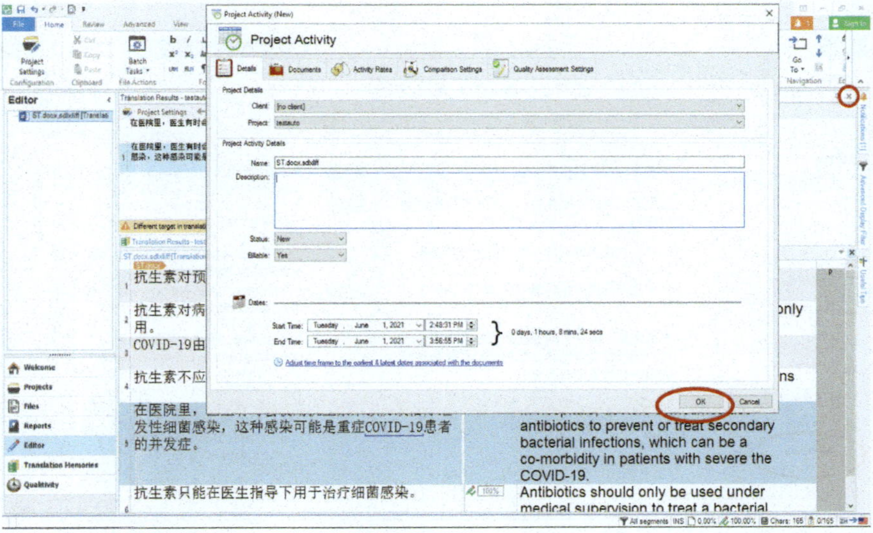

Fig. 3 Saving Qualitivity activity in Trados Studio

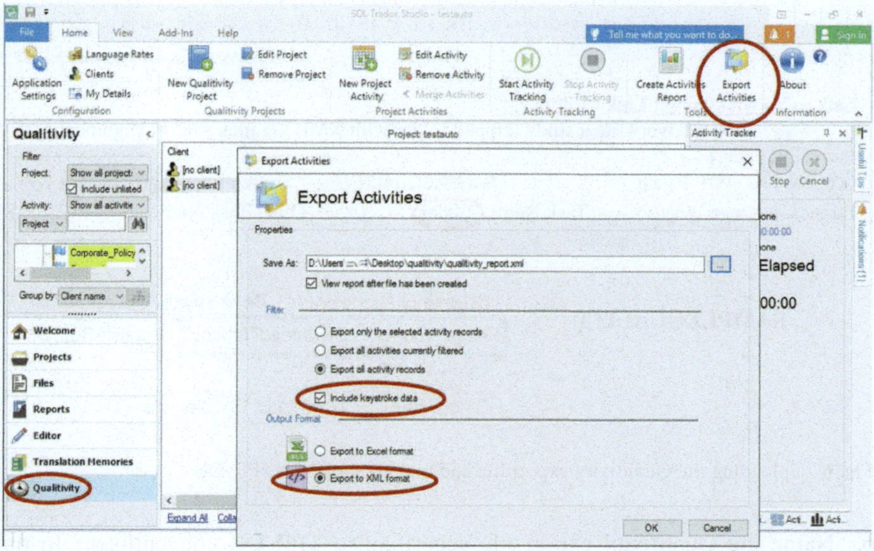

Fig. 4 User interface of the "export activities" functionality for Qualitivity plugin in Trados Studio

3 The CRITT TPR-DB

The collected data from the Qualitivity plug-in can be integrated into and processed with the CRITT TPR-DB. The CRITT TPR-DB includes a toolkit to maintain and process the raw logging data to extract and visualise translation process and product features, and to generate various fine-grained summary tables of the logged data (Carl et al. 2016). The CRITT TPR-DB has a web interface through which raw and processed data can be uploaded and a toolkit for feature extraction (e.g., fixation count and duration, gaze path, word-wise and segment-wise production duration, etc.). The procedures for data synchronisation and feature calculation are illustrated in the following Fig. 5.

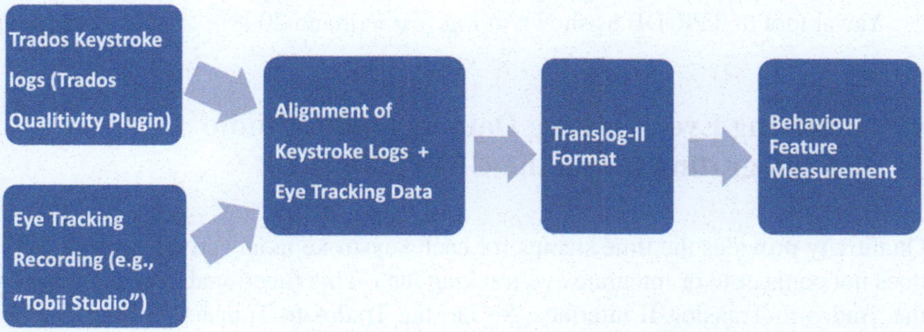

Fig. 5 CRITT TPR-DB procedures for data synchronisation and feature calculation

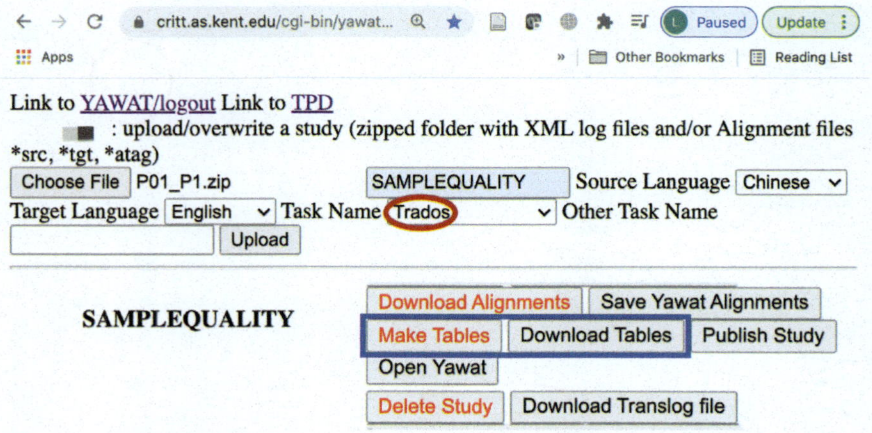

Fig. 6 Uploading the Qualitivity export file and making tables on TPR-DB

6. Name the Qualitivity export file according to TPR-DB conventions.[2] In this sample experiment named "SAMPLEQUALITY", we would save the file name as "P01_P1.xml". Zip the XML file with the same file name and upload to the TPR-DB as shown in the following screenshot Fig. 6. Select the correct source language and target language, and then select the task name as "Trados" before hitting "Upload". Click the "Make Tables" button and then click "Download Tables" or use the TPR-DB server for further analysis.
7. After conversion to the TPR-DB framework, you can gain multiple features regarding the process of post-editing through the implemented SG table, including Number of times the segment was edited ("Nedit"), production duration of each TT segment ("Dur"), the number of keystrokes which insert text per segment ("Ins"), the number of keystrokes which delete text per segment ("Del"), Word translation entropy ("HTra"), Entropy of the Cross value ("HCross") (Carl and Schaeffer 2017). Part of the features in SG tables can be seen in the following data frame in Table 1.
8. You can also investigate tokenised TPR features through the ST table and TT table on TPR-DB after aligning the source text and target text at a word level in Yawat tool of TPR-DB as shown in Fig. 7 (Germann 2008).

4 Gathering Eye-Tracking Data in Trados Studio and Integrating It into the CRITT TPR-DB

Qualitivity provides the time stamps for each keystroke in an active segment, but it does not connect to or integrate eye-tracking data. This functionality is provided by the Trados-to-Translog-II interface. So far, the Trados-to-Translog-II interface has

[2] See https://sites.google.com/site/centretranslationinnovation/tpr-db/uploading.

Table 1 Part of the features for *SG table in TPR-DB

Id	STseg	TTseg	Study	Session	SL	TL	Task	Text	Part	Nedit	Dur	FDur	PreGap	TG300	TD300	TB300
1	1	1	SAMPLEQUALITY	P01_P1	zh	en	P	1	P01	1	12,572	12,572	10,324	2248	0	1
2	2	2	SAMPLEQUALITY	P01_P1	zh	en	P	1	P01	1	13,736	13,736	8686	5049	1	4
3	3	3	SAMPLEQUALITY	P01_P1	zh	en	P	1	P01	1	15,429	15,429	14,313	611	505	1
4	4	4	SAMPLEQUALITY	P01_P1	zh	en	P	1	P01	0	0	0	0	0	0	0
5	5	5	SAMPLEQUALITY	P01_P1	zh	en	P	1	P01	1	43,895	43,895	27,786	15,191	918	5
6	6	6	SAMPLEQUALITY	P01_P1	zh	en	P	1	P01	1	14,396	43,895	11,244	2382	770	4

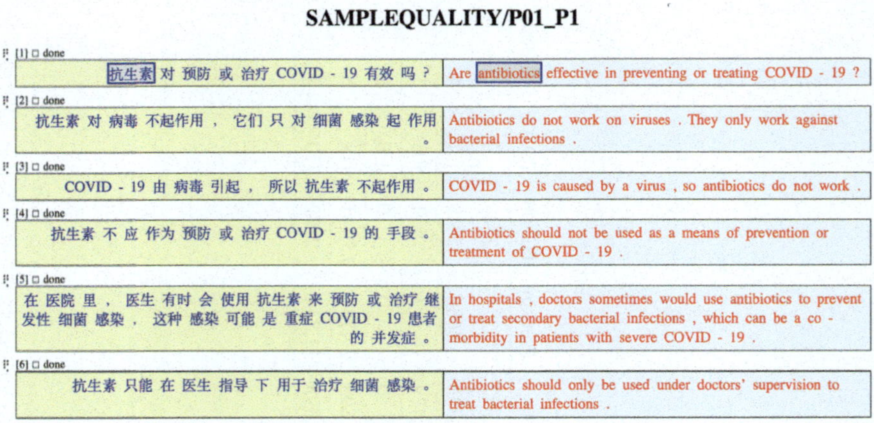

Fig. 7 Yawat alignment tool on TPR-DB

been developed for eye-trackers from Tobii, Eyelink, and GazePoint. In this section, we describe an experiment using Tobii TX300 and Tobii Studio, version 3.3.2.[3] The procedures for collecting gaze data in a Trados-to-Translog-II post-editing experiment would be as follows:

1. Create and name a new project on the eye-tracking programme. In this case as shown in the Fig. 8, we create an empty project entitled "Trados2Translog" and save the project folder under a certain directory on our local computer. Once done with this procedure, click "Next".
2. Add a new test to the current project. As shown in Fig. 9, we create the test for the first participant in our experiment by inputting the test name of "Part01". Once done with this procedure, click "Create".
3. Connect your computer to the eye-tracker. Under a proper internet environment, your computer will be automatically detected by the eye-tracker. Click on the code of the eye-tracker on the software interface of the eye-tracker and then click on "Connect to Eyetracker" (Fig. 10).
4. Make sure your participant is sitting comfortably and in a stable way at a distance of approximately 65 cm from the eye-tracker. Make sure the participant's eyes are in the middle of the black screen and the rectangle on the bottom of the black screen is green and click "Run Calibration". The calibration task usually involves following five target points on the screen with eyes. One target point would look like a red circle. If the calibration is successful, all the target points will be crossed by green and red lines. In this case, click "OK" to continue. If the calibration is unsuccessful, the interface will show some error message, such as "Not enough data to create a calibration" or the green and red lines will be quite long and scattered. In this case, redo the calibration process until it is successful.

[3] See manual and specifications of Tobii 300X at https://www.tobiipro.com/product-listing/tobii-pro-tx300/; Tobii studio: https://www.tobiipro.com/learn-and-support/learn/steps-in-an-eye-tracking-study/setup/installing-tobii-studio/.

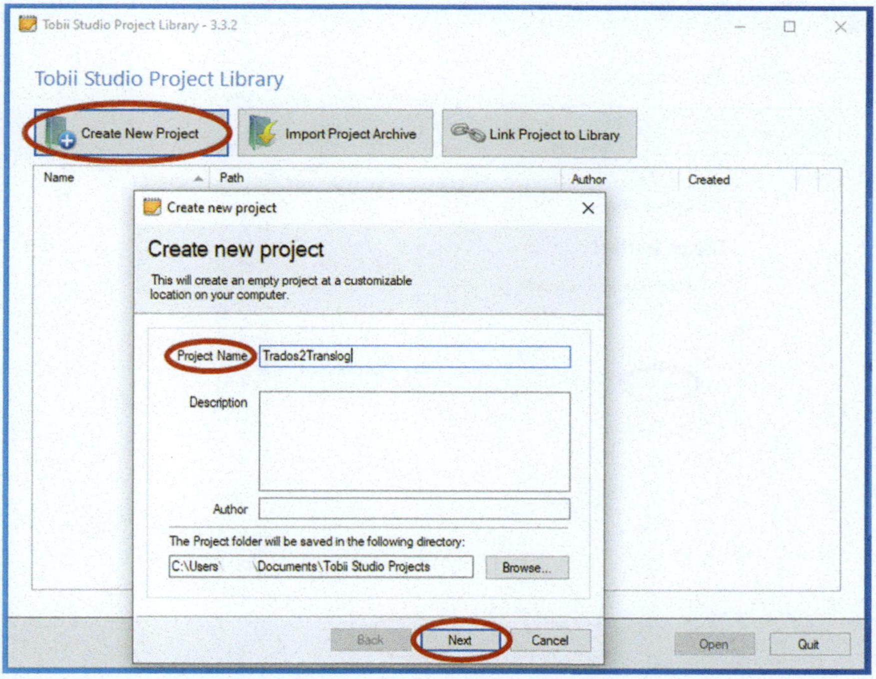

Fig. 8 Creating a new project in Tobii Studio

Once the participant finishes the calibration process, the eye-tracker recording will be ready. Click on "Screen Rec" to start recording the eye movement on the screen.

5. Open Trados Studio 2019. Click the "Projects" navigation button and set up a new post-editing project for one participant. The following procedures regarding the setup of a post-editing experiment in Trados are identical to the related descriptions in Sect. 2. After the participant finishes the post-editing experiment in Trados, make sure to save and export the Qualitivity activity in an XML format in Trados before exiting the workbench of Trados.
6. Stop screen recording by clicking certain hotkeys according to the manual of the eye-tracker (in Tobii TX300's environment, the default exiting key is ESC). Go back to Tobii Studio. Replay the eye-tracking video and manually select AOIs such as ST area and TT area, as shown in Fig. 11. Make Sure to activate these AOIs before exporting the eye-tracking data. For instance, in this sample experiment, we named two AOIs, i.e., the first AOI on the left is called "ST", and the second AOI on the right is called "TT".
7. Click the tablet of "Data Export" and select the columns for the data export (e.g., timestamp data, recording event data, gaze event data and AOI activity information, gaze tracking data, eye-tracking data, etc.) in Tobii Studio. Export the eye-tracking data collected by Tobii to a TSV format as shown in Fig. 12.

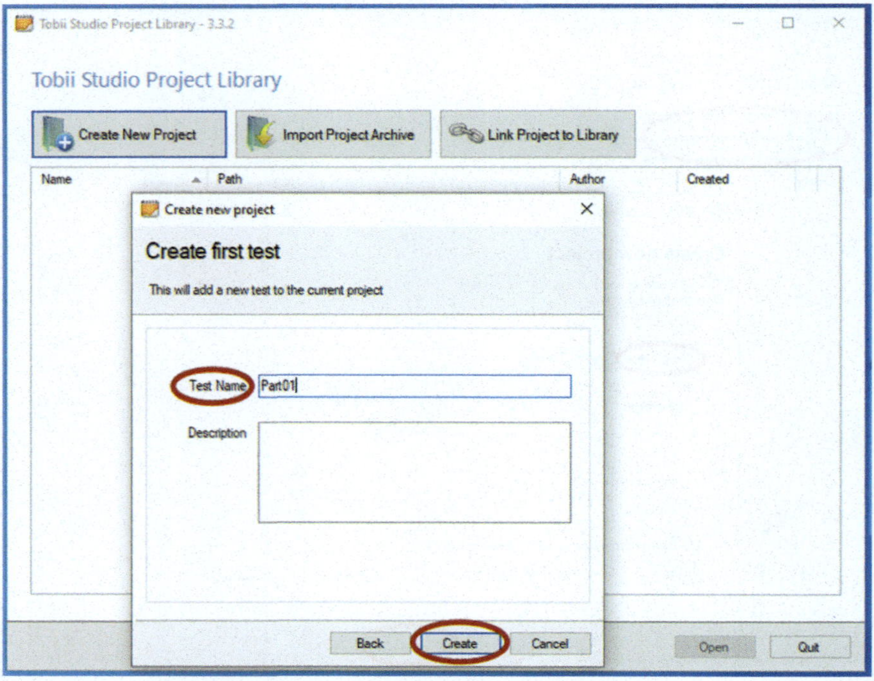

Fig. 9 Adding a new test to the current project in Tobii Studio

8. Extract at least the following columns from the data export of the eye-trackers as shown in the following Table 2. The TSV file from the eye-tracker software and the XML file from the Qualitivity plug-in should then be zipped and uploaded to the TPR-DB for further analysis.

5 An Example of Using Trados Studio to Conduct Remote Post-editing Experiments

We conducted a pilot study to test a use-case of the tool: remote post-editing experiments. Although it later proved unsuccessful, the idea was to show participants how to install and use Qualitivity, send them a Trados Studio package, have them translate the files within the package, and then export a Qualitivity keystroke log file for each of the files in the Trados package—each file representing one text production session. Sourcing participants remotely presents significant advantages, allowing researchers to draw from a far more diverse and experienced pool. This pilot study, however, demonstrated severe obstacles to remotely conducted experiments, at least as we initially imagined they would be conducted. First, we found there was a lack of control over the experimental environment (specifically, the computer system the participants used). Second, we found it problematic for participants to process and

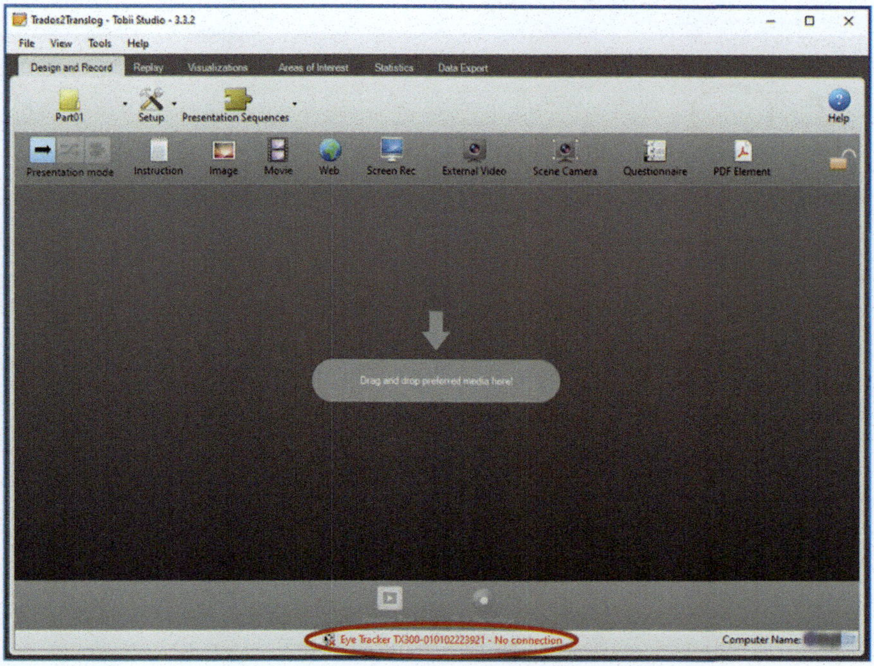

Fig. 10 Connecting local computer to the eye-tracker Tobii TX300

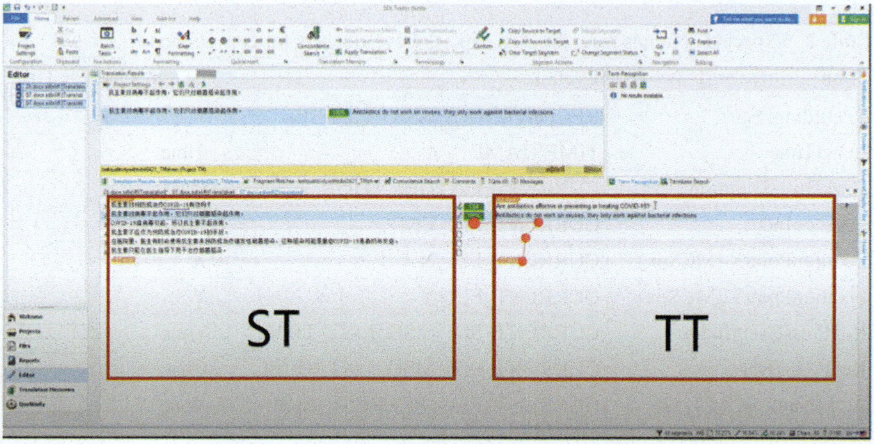

Fig. 11 Replaying eye-tracking video and selecting AOIs in Tobii Studio

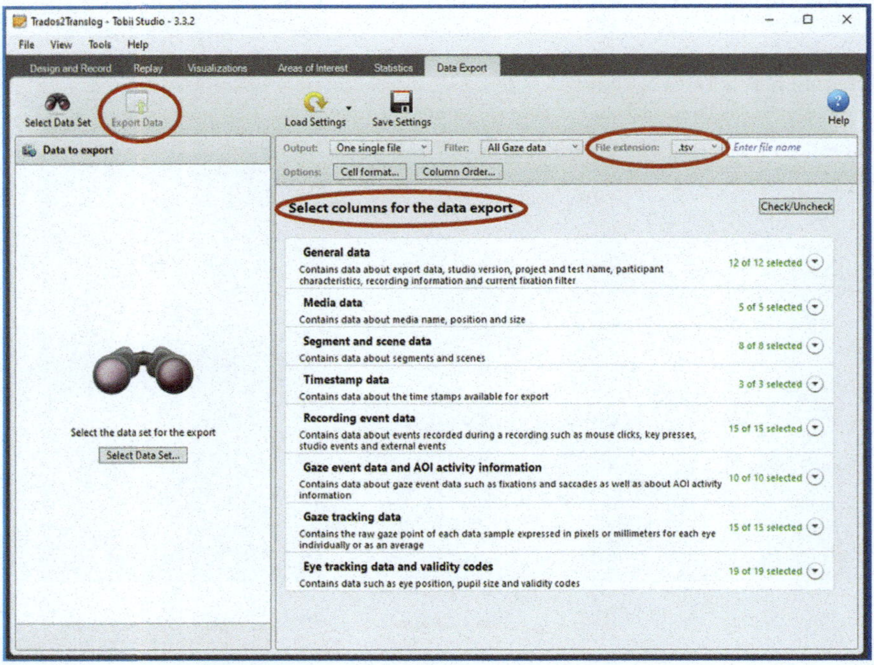

Fig. 12 Selecting columns for eye-tracking data export in Tobii Studio

Table 2 Expected columns from the data export of eye-trackers for conversion to TPR-DB

Tobii	Eyelink	Features in TPR-DB
RecordingDate	RECORDING_SESSION_LABEL	–
LocalTime	TIMESTAMP	Time
GazeEventDuration	CURRENT_FIX_DURATION	Dur
FixationIndex	CURRENT_FIX_END	Id
FixationPointX (MCSpx)	CURRENT_FIX_X	X
FixationPointY (MCSpx)	CURRENT_FIX_Y	Y
KeyPressEventIndex	CURRENT_FIX_MSG_LIST_TIME	Time
KeyPressEvent	CURRENT_FIX_MSG_LIST_TEXT	Char
AOI [Name of AOI] Hit	CURRENT_FIX_INTEREST_AREA_ LABEL	Win
PupilLeft/PupilRight	CURRENT_FIX_PUPIL	–

export data themselves. Due to these two limitations, none of the participants in our pilot study were able to successfully return intact data.

From this pilot study, we were able to condense the key steps that participants need to observe for their data to come out unscathed, and we also identified steps researchers should consider ensuring participants successfully observe these key steps. The main solution stems from using a virtual machine (VM) that participants can remotely access. The VM, an Amazon WorkSpaces[4] in our case, allows us to constrain the working environment to decrease/eliminate risks originating from user error. Here, we present a list of key steps to mitigate risks of user error; each step is accompanied by researcher actions to ensure it (indented):

- Use Trados Studio 2019, with its accompanying version of the Qualitivity plugin.
  - Control the computer on which the experiment is conducted (i.e., use a VM).
- Allow the researcher to export participant data.
  - Use a VM so the participant does not even worry about this.
- Complete each text in a single session, only opening the Trados Studio editor once per text and closing the editor before opening the next text.
  - Clearly instruct participants to only open each text in the editor once and to close it before moving on to the next text.
  - Do not instruct participants to make sure all segments have been confirmed. This is unnecessary and counterproductive since it increases the likelihood they will re-open the text in the editor after they have already finished and closed it.
- Activate every single segment in each text.
  - All the following actions apply to experiments using a translation memory (TM). TMs can be used as such or to simulate MT output. It is advisable to populate a TM beforehand with the desired MT output because it will ensure that all participants receive the exact same MT output (this cannot be guaranteed when using an API connection).

    Verify that the auto-propagate feature is disabled in the Trados Studio global "Options" ("Auto-propagation" sub-menu within the "Editor" menu).

    Verify that the "Perform lookup when active segment changes" and "Apply best match after successful lookup" settings are both enabled and that the "Confirm segment after applying an exact match" and "Enable LookAhead" settings have both been disabled (Trados Studio global "Options", "Automation" sub-menu within the "Editor" menu).

    Make sure the project settings do not pre-translate the texts (Project Settings, "All Language Pairs" menu, "Batch Processing" sub-menu). These settings ensure that TM matches automatically populate but only for the active segment.

    Make sure all specific penalties for the TM are set to zero (Project Settings, "All Language Pairs" menu, "Translation Memory and Automated Translation" sub-menu, "Penalties" sub-menu).

---

[4] See https://aws.amazon.com/workspaces/all-inclusive/.

Apply a global penalty of 1 point to the TM so that segments from the TM are not automatically confirmed.
Deselect the "Update" tick box in the TM settings so that the next participant will receive the same output from the TM.

- Click "Ok" on the Qualitivity dialog box that pops up after closing the Trados Studio editor.
  - Clearly instruct participants that they must do this; otherwise, it will result in total data loss.

Though this small-scale pilot study did not produce any actionable data, it laid the groundwork for future studies by discovering the key steps that needed to be taken. It is hoped that researchers will be able to implement these key steps as they collect TPR data in Trados Studio using the Qualitivity plugin.

After the pilot study, we conducted a more extensive PEMT study, implementing all of the key steps described above. This study had 42 participants total (Gilbert 2022). We threw out one participant's data because they did not pass the quality control segment we placed in one of the texts (we included one segment of intentionally awful "MT" output; if participants neglected to edit it, we deduced they were either incompetent or were not paying adequate attention to the task). We had to throw out an additional five participants' data because they had not observed the key steps, rendering their data unusable. Therefore, we were left with 36 participants with intact data.

This PEMT study was designed to test a feature that highlights potentially erroneous MT output for post-editors (Gilbert 2022). The question we sought to address was: will highlighting MT errors reduce effort for post-editors? Additionally, would the highlighting feature be accepted by post-editors? The 106 PEMT sessions that were the results of the 36 participants who followed the key steps were successfully integrated into the CRITT TPR-DB, demonstrating the technical feasibility of conducting remote post-editing experiments in Trados Studio on a VM and integrating the resulting data into the CRITT TPR-DB.

An analysis of the keystroke logging data showed that, although text production sessions with the highlighting feature took longer than those without the feature, there was no statistically significant difference between the experimental conditions (Gilbert 2022). In short, adding highlighting to the Trados Studio user interface actually ended up negatively impacting post-editors' productivity. Despite this negative result, post-surveys indicated that a slim majority of participants viewed the feature positively (ibid). This shows that process data may not be the only aspect to consider when introducing new features to CAT tools and that similar user interface features should include the ability to be toggled on or off, allowing linguists to decide whether or not to use them.

Although the user-interface highlighting feature this study tested was not found to increase PEMT productivity, the data revealed an interesting result with regard to the degree to which a large number of post-editors agree on which words in high-quality NMT output need to be edited: post-editors largely don't agree (ibid). The three texts that each participant post-edited add up to a total 1080 tokens; only 35 of these 1080 tokens were post-edited by 20 or more of the 36 post-editors (ibid). More studies in an ecologically valid translation/PEMT environment—such as Trados Studio—should be conducted to see if this pattern is replicable, including with other language pairs.

6 A Post-editing Behaviour Study Including Eye-Tracking Data

This section describes a feasibility study testing how keystroke data from Trados Studio can be integrated eye-tracking data using the Trados-to-Translog-II conversion tool. Heilmann (2021) recruited 3 participants in this pilot study, aiming to visualise translation behaviour measures with dimensionality reduction techniques, i.e., Principal Component Analysis (PCA). All the three participants conducted from-scratch translation tasks and two participants did post-editing tasks. The STs of the experiment were extracted from popular science news texts, and we gained altogether 37 segments of TTs in this study. Each segment was characterised by a complex set of translation behaviour.

We used PCA in this research to reveal major patterns of translation behaviour in the dataset. The first principal component (PC1) is a linear combination of behaviour predictor variables, which captures the largest variance in the dataset, whereas the second principal component (PC2) captures the second largest variance in the dataset. We looked into the reading behaviour of the translators at the segment level, including switches between ST and TT, revisiting of earlier segments, reading duration of ST and TT, parallel reading and typing, number of fixations in both ST and TT, average fixation duration, and the distance covered during the translation, etc.

We also investigated translator's typing behaviour in this study, including number of production units (PU), translation duration, and average pause length. Concerning the reading duration and typing duration at the segment level, our results in PC1 show idiosyncrasies of different translators' behaviour patterns and our results of PC2 reveal that post-edited segments are clearly separated from from-scratch translation by the behaviour of target-language focused revision.

We can see from this small-scale empirical research that identification of translation behaviour with the use of the CAT tools such as Trados Studio is feasible. TPR-DB tools such as the Trados-to-Translog-II interface can be successfully used to synchronise keystroke and gaze data from text production sessions into various data tables. Using the multiple different data tables the CRITT TPR-DB offers, researchers can examine sets of translation behaviour, particularly in combination of PCA, at many distinct levels of granularity, such as the level of the text, the segment, the alignment group, the source word, the target word, or even the keystroke. We are also able to expand the features with customised algorithms or open-source NLP packages to conduct further analysis regarding the translation effect, such as automatic translation assessment tools (e.g., BLEU and COMET) (Zou et al. 2021, 2022a, b), and linguistic complexity metrics (e.g., LingX[5]) (Zou et al. 2021, 2022a, b).

7 Using Gathered Parallel Corpus Data as a Pedagogical Tool

Whilst the CRITT TPR-DB is focused on research applications, its large collections of parallel corpora could have useful pedagogical applications. For example, students can examine parallel corpora already available in the TPR-DB, or they can examine their own process data after they have recorded a text production session in Trados Studio. Additionally, instructors can prepare purpose-built text-pairs (small, focused bilingual corpora) for student analysis using the Yawat tool (Germann 2008).

The TPR-DB houses a variety of parallel corpora. It has a diversity of language pairs (English to Danish, Dutch, German, Spanish, Chinese, Japanese, Hindi, Arabic; Portuguese, Chinese, Danish, German, Spanish, Estonian to English; Chinese to Portuguese; and Polish to French) and also includes many different modalities (from-scratch translation, post-editing, sight translation, simultaneous interpreting, and various monolingual tasks). The vast majority of studies housed by the CRITT TPR-DB are also word-aligned, and these word alignments can be visualised within the web-based Yawat environment (Germann 2008). Word-aligned source-target text pairs in the Yawat interface make it easy for students to observe professional translations and learn from novel translation solutions. For example, Fig. 13 shows how English "ice cream company" has been rendered as a single word, "heladería", in Spanish. If the user hovers their mouse over other source or target words, their corresponding translation will be highlighted in the other language. Students could be given specific tasks to complete as they observe translations in the database. For instance, they could be directed to identify non-literal or surprising translation solutions.

Yawat also makes it possible to annotate source/target words or phrases with error categories or other desired annotation schema. Students can visualise existing annotations for parallel corpora that include them, or they could also be assigned existing source-target pairs of texts to annotate themselves. Figure 14 shows the CRITT TPR-DB's default error annotation schema (based on Multidimensional Quality Metrics (MQM)), and Fig. 15 shows existing error annotations for a text that was translated from English to Spanish. The purple colour denotes a mistranslation ("Hunter-gatherer societies" was rendered as "Los agricultores [the farmers]"), whilst the pink colour on the source side indicates an omission ("hierarchical social structures" was rendered as "estructuras de jerarquía [hierarchical structures]"). If a word were highlighted in red on the target side, it would be an addition. Only unaligned words can be annotated as additions or omissions.

[5] See https://github.com/ContentSide/lingx.

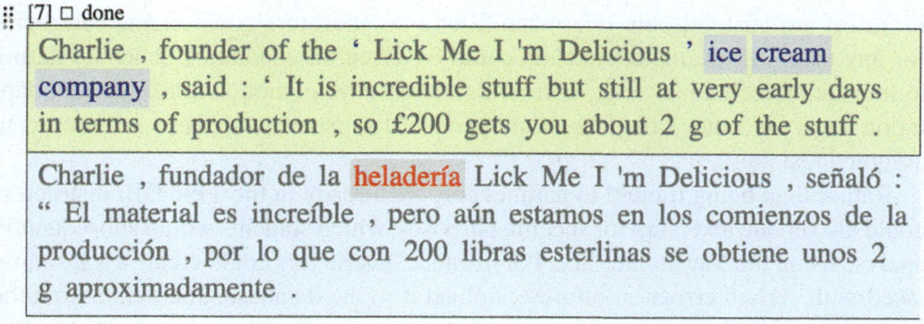

Fig. 13 Visualising word-aligned CRITT TPR-DB parallel corpora using Yawat (CRITT TPR-DB Study DG22, Participant 18, Text 14)

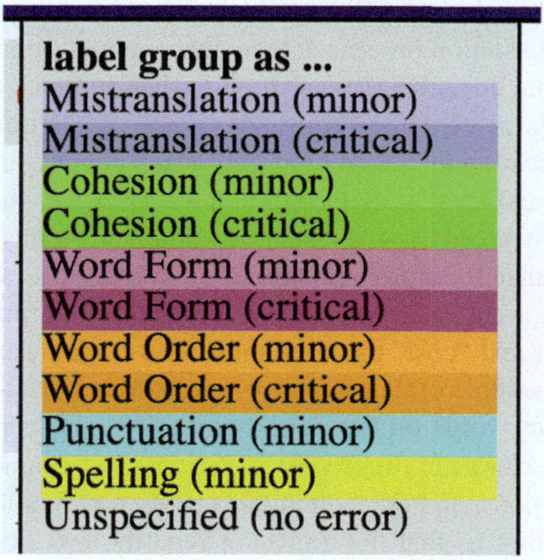

Fig. 14 CRITT TPR-DB default error annotation schema

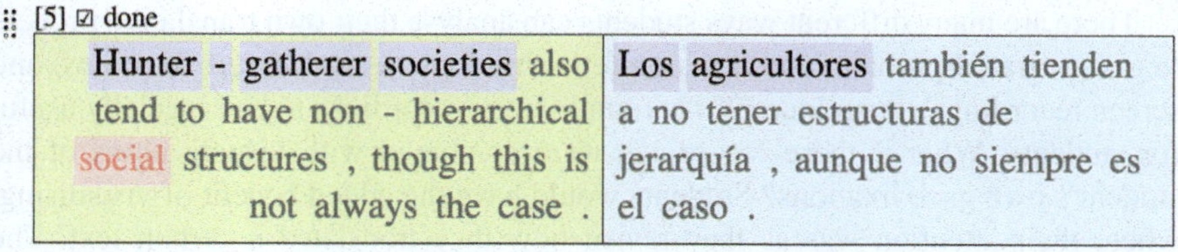

Fig. 15 Visualising existing error annotations on the CRITT TPR-DB (CRITT TPR-DB Study BML12_re, Participant 17, Text 6)

It could be helpful for students to use these colour-codings to observe errors in translation first and then to use the annotation schema themselves to error-annotate text pairs. Such an activity would aid students in developing skills to judge translations on a granular and non-holistic level. Additionally, the annotation categories in Yawat are customisable (Germann 2008), so educators could create categories for any task. For example, students could be given an annotation schema relating to translation techniques (e.g., borrowing, transposition, modulation, cultural adaptation, etc.), and they could be asked to identify instances of these techniques in assigned text pairs.

Rather than being limited to parallel corpora already in the TPR-DB, instructors could also create text pairs for specific purposes, which students would subsequently analyse using the Yawat interface. For instance, instructors could create a translation laced with certain errors on purpose, upload it to the database, and then assign the text pair to students so they can annotate it with error categories. As educators adapt annotation schemas, the same concept could be applied to other areas of focus, such as terminology or translation techniques. These focused, "mini" corpora would give students targeted practice with key concepts.

Finally, students working in Trados Studio could record, export, upload, and analyse their own translation process data to enhance metacognition. The first level of metacognition from this process would come when students word-align their own translations. As they match source words or phrases with equivalents in the target text, students are forced to reflect on their translation choices and analyse them on a granular level. The next level of metacognition would come when they analyse their own process data. We would like to present two simple ways this could be done. First, students could visualise their process data from a more general view using what is called a progression graph. Second, students could use their process data to zero in on which sections of text "cost" them the most time or gaze activity.

Figure 16 is an example of a progression graph. The Y axis shows source words on the left and target words on the right, and the X axis represents milliseconds. In the graph, black characters represent insertions, red characters deletions, blue dots source text fixations, and green diamonds target text fixations. The progression graph in Fig. 16 shows us that the translator read the text first, then translated it, and then revised. This gives us a quick view of the phases a translator goes through to accomplish their work and would be a good graphical representation to present students to think about their own processes.

There are many different ways students can analyse their own translation process (e.g., integrated problem and decision reporting logs, recorded verbalisations, and screen recordings). Angelone (2013) found screen recordings to be especially useful for students. What if these recordings were combined with visualisations of the student's own gaze fixations? Students would have the added benefit of visualising, where their attention was as they revisit how they translated a certain text. The study illustrated in Sect. 6 would effectively create such screen recordings on top of which fixations could be visualised so learners could analyse their own translation or post-editing processes.

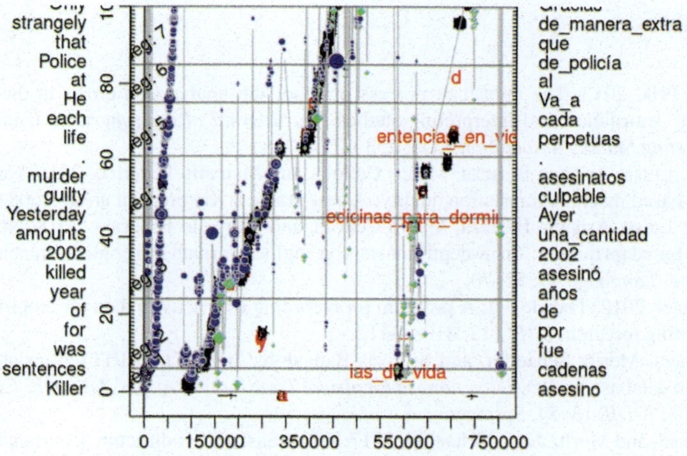

Fig. 16 Example of a progression graph of translation process data

8 Conclusion

The Qualitivity plugin for Trados Studio produces logs of keystrokes which provide segment-wise translation process information regarding both the temporal and technical effort in post-editing or translation activities. It tracks both the production time and keystroke data on a granular word level, i.e., the time spent post-editing each word, and every single change made to the words in each segment. Gaze data can be collected within Trados Studio on a segment level.

As Qualitivity works in the background of Trados Studio without any interaction from the participants whilst they are translating, it is also possible for researchers to organise the post-editing experiments remotely, for instance, within a virtual Windows machine such as Amazon WorkSpaces. This is particularly helpful for those researchers working with language pairs of lesser diffusion and scholars with difficulty finding enough subjects or subjects with sufficient translation competence.

To sum up, this paper introduces a Trados-to-Translog-II conversion tool. This tool enables scholars, student translators, and translation instructors to track different subjects' translation behaviour, increase their awareness of productivity, and characterise their translation styles. The keylogging and eye-tracking data collected and processed via the Qualitivity plugin for Trados Studio, CRITT TPR-DB, and Translog-II will provide us with detailed information about effort and behaviour in post-editing/translation experiments. A large amount of translation process data already available in the CRITT TPR-DB can be compared to data collected via the newly implemented Trados-to-Translog-II interface. The integration of eye-tracking data (e.g., Trados, Eyelink, Gazepoint) can be used by other translation scholars to investigate various research questions.

References

Angelone, Erik. 2013. The impact of process protocol self-analysis on errors in the translation product. Translation and interpreting studies. *The Journal of the American Translation and Interpreting Studies Association* 8 (2): 253–271.

Bentivogli, Luisa, Arianna Bisazza, Mauro Cettolo, and Marcello Federico. 2016. Neural versus phrase-based machine translation quality: A case study. arXiv preprint arXiv:1608.04631.

Bentivogli, Luisa, Arianna Bisazza, Mauro Cettolo, and Marcello Federico. 2018. Neural versus phrase-based mt quality: An in-depth analysis on english–german and english–french. *Computer Speech & Language* 49: 52–70.

Carl, Michael. 2012. Translog-II: A program for recording user activity data for empirical reading and writing research. *LREC* 12: 4108–4112.

Carl, Michael, Moritz Schaeffer, and Srinivas Bangalore. 2016. The CRITT translation process research database. In *New Directions in Empirical Translation Process Research: Exploring the CRITT TPR-DB*, 13–54. Springer.

Carl, Michael, and Moritz Jonas Schaeffer. 2017. Why translation is difficult: A corpus-based study of non-literality in post-editing and from-scratch translation. *HERMES-Journal of Language and Communication in Business* 56: 43–57.

Chang, Chieh Ying. 2009. *Testing Applicability of Eye-Tracking and fMRI to Translation and Interpreting Studies: An Investigation into Directionality.*

Dragsted, Barbara. 2010. Coordination of reading and writing processes in translation. *Translation and Cognition* 15: 41.

Duchowski, Andrew. 2003. *Eye Tracking Methodology: Theory and Practice*. London: Springer.

Elming, Jakob, and Laura Winther Balling. 2014. Investigating user behaviour in post-editing and translation using the CASMACAT. In *Post-editing of machine translation: Processes and Applications*, 147.

Federico, Marcello, Nicola Bertoldi, Mauro Cettolo, Matteo Negri, Marco Turchi, Marco Trombetti, Alessandro Cattelan, et al. 2014. The MateCat tool. *COLING (Demos)*, 129–132.

Germann, Ulrich. 2008. Yawat: yet another word alignment tool. In *Proceedings of the ACL-08: HLT Demo Session*, 20–23.

Gilbert, Devin. 2022. *Directing Post-Editors' Attention to Machine Translation Output that Needs Editing through an Enhanced User Interface: Viability and Automatic Application via a Word-level Translation Accuracy Indicator.*. Doctoral Dissertation, Kent State University. OhioLINK Electronic Theses and Dissertations Center. http://rave.ohiolink.edu/etdc/view?acc_num=kent1657213218346773.

Hansen-Schirra, Silvia. 2017. EEG and universal language processing in translation. In *The Handbook of Translation and Cognition*, 232–247.

Heilmann, Arndt. 2021. Translator activity during computer assisted translation. In *Book of Abstracts of the Applied Linguistics and Professional Practice Conference*, 15.

Hvelplund, Kristian Tangsgaard. 2011. *Allocation of Cognitive Resources in Translation: An Eye-Tracking and Key-Logging Study*. Frederiksberg: Copenhagen Business School (CBS).

Jakobsen, Arnt Lykke. 1999. Logging Target Text Production with Translog. *Copenhagen Studies in Language* 24: 9–20.

Jia, Yanfang, Michael Carl, and Xiangling Wang. 2019. How does the post-editing of neural machine translation compare with from-scratch translation? A product and process study. *The Journal of Specialised Translation* 31 (1): 60–86.

Just, Marcel A., and Patricia A. Carpenter. 1980. A theory of reading: From eye fixations to comprehension. *Psychological Review* 87 (4): 329.

Krings, Hans P. 2001. Repairing texts: Empirical investigations of machine translation post-editing processes. In *Translation Studies*, vol. 5. Kent: Kent State University Press.

Lacruz, Isabel. 2017. Cognitive effort in translation, editing, and post-editing. In *The Handbook of Translation and Cognition*, ed. John W. Schwieter and Aline Ferreira, 386–401. Malden: Wiley.

Lacruz, Isabel, Michael Denkowski, and Alon Lavie. 2014. Cognitive demand and cognitive effort in post-editing. In *Proceedings of the 11th Conference of the Association for Machine Translation in the Americas*, 73–84.

Läubli, Samuel, Sheila Castilho, Graham Neubig, Rico Sennrich, Qinlan Shen, and Antonio Toral. 2020. A set of recommendations for assessing human–machine parity in language translation. *Journal of Artificial Intelligence Research* 67: 653–672.

Leijten, Mariëlle, and Luuk Van Waes. 2013. Keystroke logging in writing research: Using Inputlog to analyze and visualize writing processes. *Written Communication* 30 (3): 358–392.

Mellinger, Christopher D., and Thomas A. Hanson. 2022. Considerations of ecological validity in cognitive translation and interpreting studies. *Translation, Cognition and Behaviour* 5 (1): 1–26.

Moorkens, Joss. 2018. Eye tracking as a measure of cognitive effort for post-editing of machine translation. *Eye Tracking and Multidisciplinary Studies on Translation*, 55–69.

O'Brien, Sharon. 2006. Eye-tracking and translation memory matches. *Perspectives-Studies in Translation Theory and Practice* 14 (3): 185–205.

O'Brien, Sharon. 2009. Eye tracking in translation process research: Methodological challenges and solutions. *Methodology, Technology and Innovation in Translation Process Research* 38: 251–266.

Snover, Matthew, Bonnie Dorr, Richard Schwartz, Linnea Micciulla, and John Makhoul. 2006. A study of translation edit rate with targeted human annotation. In *Proceedings of the 7th Conference of the Association for Machine Translation in the Americas: Technical Papers*, 223–231.

Sun, Sanjun. 2019. Measuring difficulty in translation and post-editing: A review. *Researching Cognitive Processes of Translation*, by Defeng Li, Victoria Lai Cheng Lei, and Yuanjian He, 139–168. Springer.

Toral, Antonio, and Víctor M. Sánchez-Cartagena. 2017. A multifaceted evaluation of neural versus phrase-based machine translation for 9 language directions. In *Proceedings of the 15th Conference of the European Chapter of the Association for Computational Linguistics*, vol. 1, Long Papers (Valencia), 1063–1073.

Toral, Antonio, Martijn Wieling, and Andy Way. 2018. Post-editing effort of a novel with statistical and neural machine translation. *Frontiers in Digital Humanities* 9.

Vieira, Lucas Nunes. 2016. Cognitive effort in post-editing of machine translation: Evidence from eye movements, subjective ratings, and think-aloud protocols. PhD diss., Newcastle University, 2016.

Vieira, Lucas Nunes, Elisa Alonso, and Lindsay Bywood. 2019. Introduction: Post-editing in practice–Process, product and networks. *The Journal of Specialised Translation* 31: 2–13.

Vieira, Lucas Nunes, Valentina Ragni, and Elisa Alonso. 2021. Translator autonomy in the age of behavioural data. *Translation, Cognition and Behaviour* 4 (1): 124–146.

Walker, Callum, and Federico M. Federici, eds. 2018. *Eye tracking and multidisciplinary studies on translation*, vol. 143. Amsterdam: John Benjamins Publishing Company.

Yamada, Masaru, Takanori Mizowaki, Longhui Zou, and Michael Carl. 2022. Trados-to-Translog-II: Adding gaze and qualitivity data to the CRITT TPR-DB. In *Proceedings of the 23rd Annual Conference of the European Association for Machine Translation*, 293–294.

Zou, Longhui, Michael Carl, Mehdi Mirzapour, Hélène Jacquenet, and Lucas Nunes Vieira. 2021. Ai-based syntactic complexity metrics and sight interpreting performance. In *Intelligent Human Computer Interaction: 13th International Conference, IHCI 2021, Kent, OH, USA, December 20–22, 2021, Revised Selected Papers*, 534–547. Cham: Springer International Publishing.

Zou, Longhui, and Michael Carl. 2022. Trados and the critt tpr-db: Translation process research in an ecologically valid environment. In *Model Building in Empirical Translation Studies: Proceedings of TRICKLET Conference*, 38–40.

Zou, Longhui, Ali Saeedi, and Michael Carl. 2022a. Investigating the impact of different pivot languages on translation quality. In *Proceedings of the 15th Biennial Conference of the Association for Machine Translation in the Americas* (Workshop 1: Empirical Translation Process Research), 15–28.

Zou, Longhui, Michael Carl, Masaru Yamada, and Takanori Mizowaki. 2022b. Proficiency and external aides: Impact of translation brief and search conditions on post-editing quality. In *Proceedings of the 15th Biennial Conference of the Association for Machine Translation in the Americas (Workshop 1: Empirical Translation Process Research)*, 60–74.

Longhui Zou completed her M.A. in Interpreting and Translation Studies from Wake Forest University in the US. She is currently a doctoral fellow at Kent State University. Longhui focuses on Chinese and English translation and interpreting, as well as machine translation and post-editing, translation and interpreting processes, and translation technologies. Her most recent effort intends to collect new and analyze legacy keylogging and eye tracking data to investigate translation processes and identify higher-order cognition based on behavioral patterns of monitoring activity observed in logged translation sessions.

Michael Carl is a distinguished Professor at Kent State University/USA and Director of the Center for Research and Innovation in Translation and Translation Technology (CRITT). He has studied Computational Linguistics and Communication Sciences in Berlin, Paris and Hong Kong and obtained his Ph.D. degree in Computer Sciences from the Saarland University/Germany. He has worked and published for more than 25 years in the fields on translation studies, machine translation and natural language processing. His current research interest is related to the investigation of human translation processes and interactive machine translation.

Devin Gilbert is a translation/interpreting practitioner, educator, and researcher and is currently an Assistant Professor of Spanish and Translation and Interpreting at Utah Valley University, where he leads their translation and interpreting programme. His research interests include translation/interpreting pedagogy, translation process research, and translation/interpreting technology. He enjoys developing software for the web, using code to make data analysis easier, and nerding out on all things language.

Corpora and Translation Teaching

Creating and Using "Virtual Corpora" to Extract and Analyse Domain-Specific Vocabulary at English-Corpora.org

Mark Davies

1 Introduction

One of the most challenging tasks facing translators is the task of creating a robust term bank of words and phrases for a particular domain, such as electrical engineering, endocrinology, or investments. Word lists such as the General Service List (West 1953), the New General Service List (Browne et al. 2013), or the word frequency data from the one billion word, genre-balanced Corpus of Contemporary American English (Davies 2008b) will be far too general in nature for a particular domain. These general lists will contain words like *smile*, *laugh*, *puppy*, *window*, *sweet*, *thin*, and *happily*, which will be found very rarely (if at all) in texts from technical domains. Even academically-oriented lists like the Academic Word List (Coxhead 2000) or the Academic Vocabulary List (Gardner and Davies 2013), or academic multiword expressions (Simpson-Vlach and Ellis 2010; Ackermann and Chen 2013) will likely not be specific enough, since the lists require that the words occur across multiple domains. As a result, words and phrases like *bandwidth*, *generator*, or *control system* (electrical engineering), *hormone*, *steroids*, or *glucose level* (endocrinology), or *beneficiary*, *pension*, or *venture capital* (investments) will not be found in these academic word lists or lists of academic multiword expressions.

Previous researchers have shown the value of creating corpora for these specific domains, and then extracting frequency data from these specialised corpora (see Davies 2019). The basic method is to compare the words and phrases from these specialised corpora to the general (or even general academic) language, to extract domain-specific words and phrases. This approach has been the focus of studies like Castagnoli (2006), Charles (2014), Lee and Swales (2006), Mudraya (2006), and Smith (2015). One challenge, however, is the sheer difficulty and amount of time required to create such corpora. If it takes hours, days, or even weeks to create a

M. Davies (✉)
English-Corpora.org, Springville, USA
e-mail: mark.davies@english-corpora.org

J. Pan and S. Laviosa (eds.), *Corpora and Translation Education*, New Frontiers in Translation Studies, https://doi.org/10.1007/978-981-99-6589-2_5

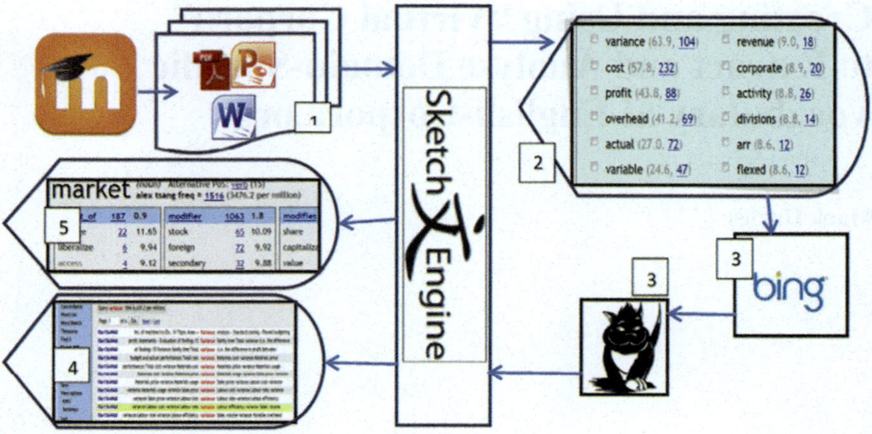

Fig. 1 BootCat with sketch engine

one or two million word corpus for a particular domain (and then even more time to enlarge or refine the corpus), this approach quickly loses its appeal for translators and other researchers.

Smith (2020) discusses how BootCat (Baroni and Bernadini 2004; Baroni et al. 2006) can be used with Sketch Engine (Kilgarrif et al. 2004) to create what he calls "do it yourself (DIY) corpora". Figure 1 provides an overview of the process.

As Smith notes, users create a small "seed" corpus composed of documents related to the particular domain (1). BootCat and Sketch Engine then generate a list of words and phrases that are much more frequent in that small corpus than in general English (2). BootCat then uses these words as a "seed" for online searches (such as with Bing) to find many more texts that contain those words and phrases. Once these texts are downloaded and annotated, they can be analysed in Sketch Engine, to find n-grams, collocates, and more (5 and 6).

In this chapter, we will discuss a method that is similar to Smith (2020). Our approach uses the corpora from English-Corpora.org (Davies 2008), which contain tens of billions of words of data from tens of millions of texts. As discussed in Sect. 2, from within the corpus interface itself, we can easily create "virtual corpora" using either words or phrases, or using metadata about the texts. We can create these corpora in just a few seconds, which is probably hundreds or even thousands of times faster than the approach in Smith (2020). In Sect. 3, we show how we can modify and organise these virtual corpora, again within the corpus interface. Section 4 shows how we can easily and quickly (1–2 s) extract keywords and even "key phrase" multiword expressions from the virtual corpora. And finally, in Sect. 5, we discuss how we can analyse the virtual corpora by searching for n-grams, collocates, and more.

2 Creating Virtual Corpora

Creating virtual corpora in the corpora from English-Corpora.org is both very quick and very easy. In the discussion that follows, we will discuss the two main ways of creating virtual corpora: via words and phrases in the texts, and by using metadata about the texts.

Before discussing virtual corpora, however, it might be helpful to briefly introduce some of the corpora that are available from English-Corpora.org, and which we will refer to throughout this chapter (Table 1).

One of the things that distinguishes English-Corpora.org from other online collections of corpora is the degree to which these corpora allow users to examine variation in English. Most other large corpora (400+ million words) are taken primarily from

Table 1 Corpora from English-Corpora.org (sizes as of Oct 2021)

Corpus	# Words	Dialect	Time period	Genre(s)
iWeb: the intelligent web-based corpus	14 billion	6 countries	2017	Web
News on the web (NOW)	17.4 billion+	20 countries	2010–yesterday	Web: news
Global web-based English (GloWbE)	1.9 billion	20 countries	2012–13	Web (incl blogs)
Wikipedia corpus	1.9 billion	(Various)	2014	Wikipedia
Hansard corpus	1.6 billion	British	1803–2005	Parliament
Coronavirus corpus	1.6 billion+	20 countries	Jan 2020–yesterday	Web: news
Corpus Contemp Am English (COCA)	1.0 billion	American	1990–2019	Balanced
Early English books online	755 million	British	1470s–1690s	(Various)
Corpus Historical Am English (COHA)	475 million	American	1820–2019	Balanced
The TV corpus	325 million	6 countries	1950–2018	TV shows
The movie corpus	200 million	6 countries	1930–2018	Movies
Corpus of US supreme court opinions	130 million	American	1790s–present	Legal opinions
Corpus of American soap operas	100 million	American	2001–2012	TV shows
TIME magazine corpus	100 million	American	1923–2006	Magazine
British national corpus (BNC)	100 million	British	1980s–1993	Balanced
Strathy corpus (Canada)	50 million	Canadian	1970s–2000s	Balanced
CORE corpus	50 million	6 countries	2014	Web

web pages, which are often just a large "blob" of texts. At English-Corpora.org, however, the corpora are designed in such a way that it is easy to examine genre-based, historical, and dialectal variation in English—to a degree that is probably not possible anywhere else (see Davies 2017, 2018). This may be part of the reason that researchers, translators, teachers, and students use English-Corpora.org more than any other collection of online corpora.

2.1 Creating Virtual Corpora Using Words and Phrases

Via the main search interface for a corpus, users can simply enter a word, phrase, or substring. Figure 2a shows the search for the phrase *nuclear power* in the 14 billion word iWeb Corpus; Fig. 2b shows the search for the word *refugees* in the 13.6 billion word NOW Corpus (here limited to texts from 1 Sep 2015 to 30 Nov 2015); and Fig. 2c is the search *endocrin** (e.g. *endocrine*, *endocrinologist*) in the 2 billion word Wikipedia Corpus.

In less than one second, the corpus finds what it thinks are the best websites (iWeb), articles (Wikipedia), or texts (all other corpora) for the search. Figure 3a–c show the matching websites, articles, or texts for these three searches. The reason that these searches are so fast is that the underlying architecture for the corpora relies on relational databases, where a lot of crucial information (frequency of words by text, etc.) is stored in the database, and simply needs to be retrieved.

By default, the corpus will find the texts where the raw frequency of the phrase, word, or substring is the greatest (as in Fig. 3). But this may favour texts or websites that are larger in size, and where there would be more features. In order to take this into account, it is also possible to sort by relevance. For example, in Fig. 4, we find those websites, where the normalised frequency of *nuclear energy* is the highest, per thousand words.

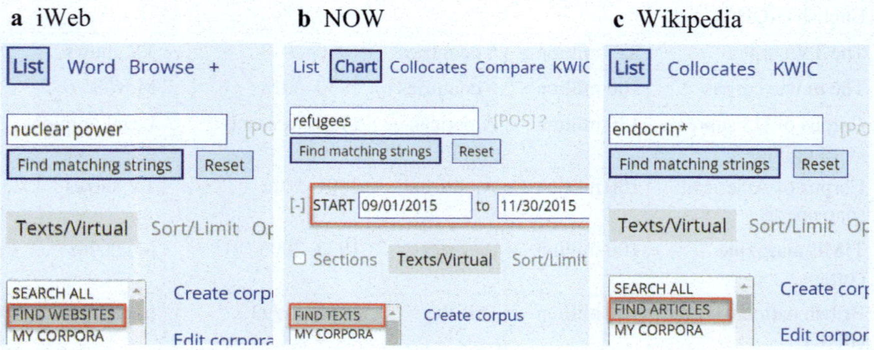

Fig. 2 Creating virtual corpora via words and phrases in texts

a

(Optional) SAVE AS: nuclear-power OR ADD TO: --SELECT-- SUBMIT RESET HELP

HELP		TEXT	# WORDS	# HITS	RELEVANCE	PER MILLION WORDS	KEYWORDS
1	☑	WORLD-NUCLEAR.ORG	1097540	2907	2,648.7		reactor, uranium, decommissioning, nuclear, radioactive, atomic, fuel, electricity
2	☑	WWW.WORLD-NUCLEAR-NEWS.ORG	496222	1911	3,851.1		reactor, nuclear, fuel, plant, electricity, construction, unit, energy
3	☑	ECOLO.ORG	345232	846	2,450.5		reactor, uranium, environmentalist, radioactive, nuclear, radiation, coal, dioxide
4	☑	NEI.ORG	179328	573	3,195.3		reactor, nuclear, electricity, plant, regulatory, fuel, energy, emission
5	☑	NRC.GOV	515077	491	953.3		licensee, radioactive, reactor, appendix, radiation, nuclear, paragraph, license
6	☑	IEER.ORG	311625	441	1,415.2		plutonium, reactor, uranium, radioactive, nuclear, waste, radiation, weapon
7	☑	NUCLEAR-ENERGY.NET	62002	376	6,064.3		fission, reactor, neutron, isotope, nuclear, uranium, radioactive, atomic

b

(Optional) SAVE AS: refugees OR ADD TO: --SELECT-- SUBMIT RESET HELP

HELP		TEXT	# WORDS	# HITS	RELEVANCE	PER MILLION WORDS
1	☑	PH 19-05-30: ABS-CBN NEWS	29327	186	6,342.3	
2	☑	AU 16-05-19: ABC ONLINE	40571	177	4,362.7	
3	☑	AU 16-05-18: ABC ONLINE	33160	165	4,975.9	
4	☑	AU 16-05-02: ABC ONLINE	44282	155	3,500.3	
5	☑	GB 20-09-29: INDEPENDENT.CO.UK	3740	135	36,096.3	
6	☑	US 21-05-17: NATURE	11093	135	12,169.8	
7	☑	AU 16-05-04: ABC ONLINE	36553	121	3,310.3	
8	☑	AU 19-10-15: LOWYINSTITUTE.ORG	9080	121	13,326.0	
9	☑	GB 17-01-25: THE INDEPENDENT	2424	111	45,792.1	

c

(Optional) SAVE AS: endocrin_ OR ADD TO: --SELECT-- SUBMIT RESET HELP

HELP		MORE SEARCHES	SEE WORD IN CONTEXT	SEE PAGE	# WORDS	# HITS	RELEVANCE	PER MILLION WORDS
1	☑		ENDOCRINE DISRUPTOR		5,739	64	11,151.8	
2	☑		HOSSEIN GHARIB		1,976	56	28,340.1	
3	☑		ENDOCRINOLOGY		2,058	50	24,295.4	
4	☑		CYRIL Y. BOWERS		2,095	42	20,047.7	
5	☑		EUROPEAN SOCIETY OF ENDOCRINOLOGY		267	25	93,633.0	
6	☑		MULTIPLE ENDOCRINE NEOPLASIA		1,291	25	19,364.8	
7	☑		ENDOCRINE SYSTEM		916	24	26,200.9	

Fig. 3 Creating corpora by word or phrase **a** iWeb **b** NOW **c** Wikipedia

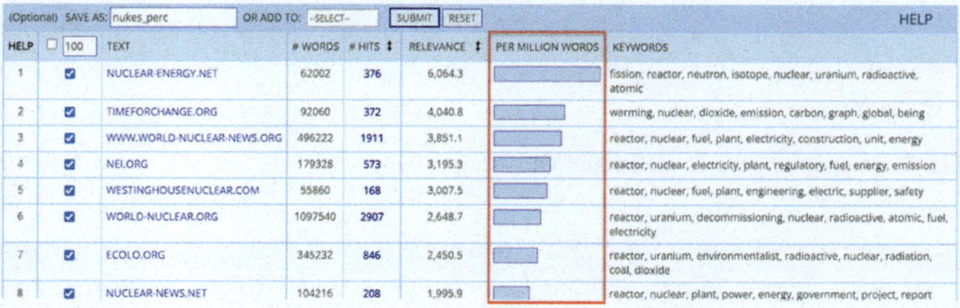

(Optional) SAVE AS: nukes_perc OR ADD TO: --SELECT-- SUBMIT RESET HELP

HELP		TEXT	# WORDS	# HITS	RELEVANCE	PER MILLION WORDS	KEYWORDS
1	☑	NUCLEAR-ENERGY.NET	62002	376	6,064.3		fission, reactor, neutron, isotope, nuclear, uranium, radioactive, atomic
2	☑	TIMEFORCHANGE.ORG	92060	372	4,040.8		warming, nuclear, dioxide, emission, carbon, graph, global, being
3	☑	WWW.WORLD-NUCLEAR-NEWS.ORG	496222	1911	3,851.1		reactor, nuclear, fuel, plant, electricity, construction, unit, energy
4	☑	NEI.ORG	179328	573	3,195.3		reactor, nuclear, electricity, plant, regulatory, fuel, energy, emission
5	☑	WESTINGHOUSENUCLEAR.COM	55860	168	3,007.5		reactor, nuclear, fuel, plant, engineering, electric, supplier, safety
6	☑	WORLD-NUCLEAR.ORG	1097540	2907	2,648.7		reactor, uranium, decommissioning, nuclear, radioactive, atomic, fuel, electricity
7	☑	ECOLO.ORG	345232	846	2,450.5		reactor, uranium, environmentalist, radioactive, nuclear, radiation, coal, dioxide
8	☑	NUCLEAR-NEWS.NET	104216	208	1,995.9		reactor, nuclear, plant, power, energy, government, project, report

Fig. 4 Creating corpora; sorting by tokens per million words

2.2 *Creating Virtual Corpora via Metadata*

A second way of creating virtual corpora is to use the metadata information for each text. For example, Fig. 5 shows a search for articles in COCA (Corpus of Contemporary American English) in financial magazine articles (with *Money* in the title of the magazine) from 1990 to 1999, where the substring *retir** (*retire*, *retiring* etc.) is the title of the article. Figure 6 shows partial results for this search.

Another example is from the 17.4 billion word NOW Corpus (News on the Web), which contains more than 29 million texts from January 2010 through the current time (May 2023, as of the writing of this chapter), and which continues to grow by about 300,000–350,000 texts and about 200 million words each month. The metadata for each text contains the date, country, source (e.g. New York Times), and title of the text, and translators and researchers can use any of this information to create a virtual corpus. For example, Fig. 7 shows a search to create a virtual corpus based on articles from *The Guardian* newspaper (in the UK/Great Britain) from September through December 2015, which contain the word *asylum*. Figure 8 shows a partial listing of the matching articles.

For some corpora, even more metadata is available. For example, Fig. 9 shows the form to create a virtual corpus from the TV Corpus (325 million words, 1950s–2010s). Users can search by series title (e.g. *Star Trek* or *Doctor Who*), genre (e.g. Sci-Fi or Drama), country, TV rating, IMDB rating (for example, to find series that

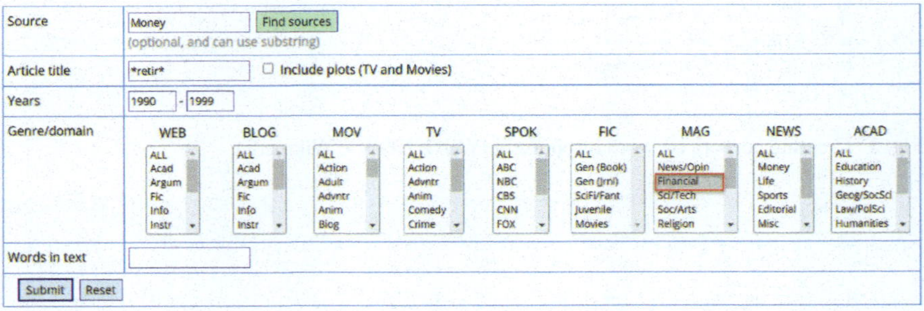

Fig. 5 Creating corpora via metadata: COCA Corpus

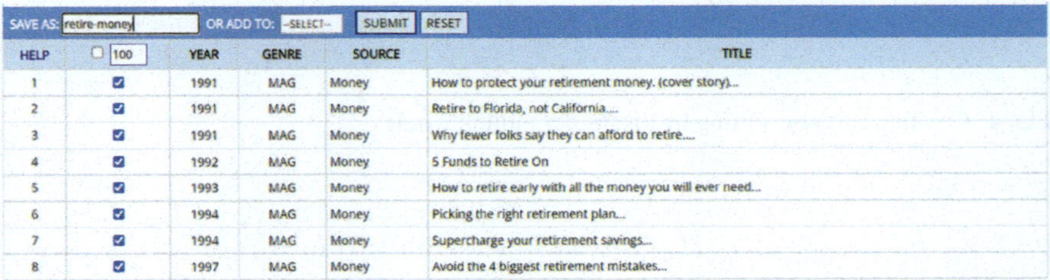

HELP	100	YEAR	GENRE	SOURCE	TITLE
1	☑	1991	MAG	Money	How to protect your retirement money. (cover story)...
2	☑	1991	MAG	Money	Retire to Florida, not California....
3	☑	1991	MAG	Money	Why fewer folks say they can afford to retire....
4	☑	1992	MAG	Money	5 Funds to Retire On
5	☑	1993	MAG	Money	How to retire early with all the money you will ever need...
6	☑	1994	MAG	Money	Picking the right retirement plan...
7	☑	1994	MAG	Money	Supercharge your retirement savings...
8	☑	1997	MAG	Money	Avoid the 4 biggest retirement mistakes...

Fig. 6 Creating corpora via metadata: results from COCA

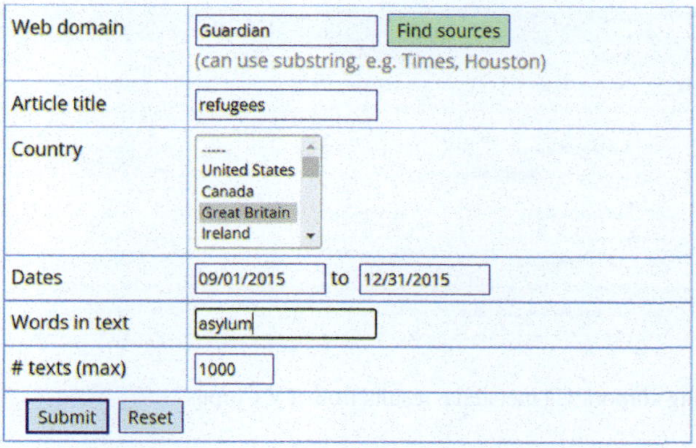

Fig. 7 Creating corpora via metadata: NOW Corpus

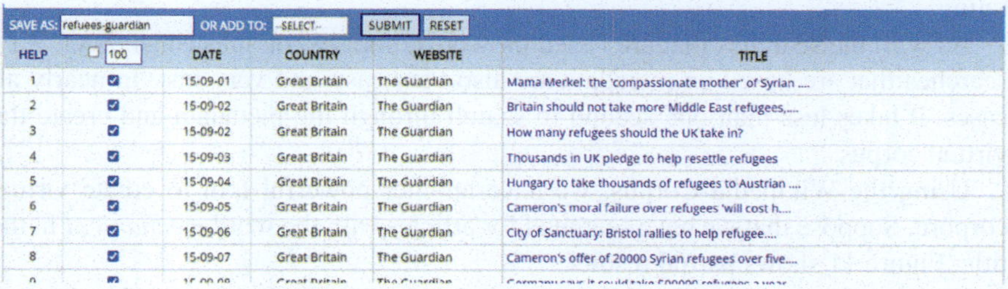

HELP	100	DATE	COUNTRY	WEBSITE	TITLE
1	☑	15-09-01	Great Britain	The Guardian	Mama Merkel: the 'compassionate mother' of Syrian ....
2	☑	15-09-02	Great Britain	The Guardian	Britain should not take more Middle East refugees,....
3	☑	15-09-02	Great Britain	The Guardian	How many refugees should the UK take in?
4	☑	15-09-03	Great Britain	The Guardian	Thousands in UK pledge to help resettle refugees
5	☑	15-09-04	Great Britain	The Guardian	Hungary to take thousands of refugees to Austrian ....
6	☑	15-09-05	Great Britain	The Guardian	Cameron's moral failure over refugees 'will cost h....
7	☑	15-09-06	Great Britain	The Guardian	City of Sanctuary: Bristol rallies to help refugee....
8	☑	15-09-07	Great Britain	The Guardian	Cameron's offer of 20000 Syrian refugees over five....

Fig. 8 Creating corpora via metadata: results from NOW

were highly or poorly rated), the plot for a specific episode (e.g. *time travel* or *gay* or *marriage*), and words in the text.

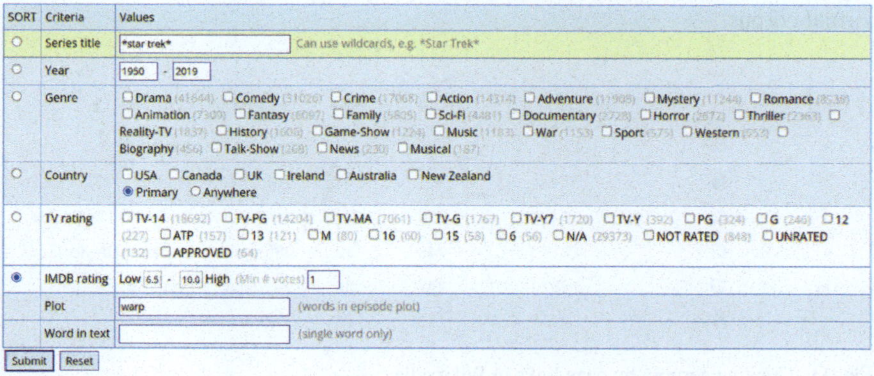

Fig. 9 Creating corpora via metadata: TV corpus

SAVE AS: starTrek-warp OR ADD TO: --SELECT-- SUBMIT RESET

HELP	100	YEAR	SERIES	EPISODE	COUNTRY	GENRE	RATING	IMDB
1	☑	1999	Star Trek: Voyager	Dark Frontier	USA	Action, Adventure, Sci-Fi	TV-PG	8.4 (743)
Members of the Voyager crew train on the holodeck for a raid on a Borg ship. Should they be successful, they will steal the Borg trans-**warp** coil in hopes of integrating the technology into ...								
2	☑	2004	Star Trek: Enterprise	Damage	USA	Action, Adventure, Drama	TV-PG	8.3 (667)
Captain Archer has been returned to Enterprise by the Xindi council awaiting a hearing. But after a heavy attack by the Reptilians, Enterprise is stranded without a functional **warp** drive. ...								
3	☑	2001	Star Trek: Enterprise	Broken Bow: Part I	USA	Action, Adventure, Drama	PG	7.6 (3267)
Enterprise, Earth's first Warp 5 vessel, embarks on a dangerous first mission: bringing back a chased Klingon to his home world of Qo'noS.								
4	☑	2003	Star Trek: Enterprise	First Flight	USA	Action, Adventure, Drama	N/A	7.6 (744)
After his close friend and colleague A.G. Robinson is killed in a mountain climbing accident, Captain Archer tells T'Pol about the Warp 2 test flight almost ten years earlier.								
5	☑	1993	Star Trek: The Next Generation	Phantasms	USA	Action, Adventure, Mystery	TV-PG	7.6 (1056)
Data deals with his new ability to experience nightmares as the Enterprise deals with the installation of a new **warp** core that won't work as expected.								

Fig. 10 Creating corpora via metadata: results from TV Corpus

Figure 10 shows just the first few results for a search of Star Trek episodes with the word *warp* in the plot description for the episode, and sorted by their IMDB rating.

As with the searches that are based on words, phrases, or substrings (Sect. 2.1), searches that are based on metadata are also very easy and very fast. In nearly all cases, it takes less than one second to search through the metadata and create the virtual corpus.

Using the Wikipedia Corpus, we have another powerful tool to create virtual corpora. Suppose that we have searched for articles with the word *mechanical* in the title. Figure 11 shows partial results.

We can click on (1) to find all articles in Wikipedia that link to the [Mechanical engineering] article (2,452 articles), or (2) to find all articles that the [Mechanical engineering] (168) article links to. This list of linked pages is both fast and powerful, because we have stored tens of millions of links in a relational database (with source article and target article/hyperlink). If we choose the second option (pages that the article [Mechanical engineering] links to), we will see a list of articles like those in Fig. 12. We simply select the articles that are of interest to us, and add them to our virtual corpus.

HELP	100	# WORDS	(1)	(2)	TITLE (SEE IN WIKIPEDIA)
1	☑	4,215	2452	168	Mechanical engineering
2	☑	3,848	313	84	Bearing (mechanical)
3	☑	918	254	37	Institution of Mechanical Engineers
4	☑	4,818	248	80	Differential (mechanical device)
5	☑	3,787	247	88	Microelectromechanical systems
6	☑	1,395	213	72	Mechanical ventilation

Fig. 11 Creating corpora by page links in Wikipedia

28	☑	3,995	1133	230	Transport
29	☑	3,829	1110	108	Logistics
30	☑	8,992	1101	86	Electric motor
31	☑	1,563	1039	97	Sound
32	☑	4,572	1014	191	Engine
33	☑	9,057	1007	289	Optics
34	☑	6,330	960	69	Heat
35	☑	6,195	932	179	Robotics
36	☑	1,488	902	76	Bit
37	☑	2,048	887	47	Software engineering
38	☑	10,284	882	121	Thermodynamics

Fig. 12 Links from *mechanical engineering* in Wikipedia

3 Organising and Refining the Virtual Corpora

The virtual corpora are not necessarily static. As Fig. 13 shows (with a [telescope] corpus in COCA), after users create a virtual corpus, users can select texts and then [DELETE] the texts, [ADD] them to another virtual corpus, or [MOVE] them to another virtual corpus.

For example, users might create virtual corpora of texts dealing with different topics in medical research (e.g. endocrinology, cancer research, neurophysiology, or cardiology), and then cut, copy, and move texts from one virtual corpus to another.

Users can also review lists of all of their virtual corpora, as in Fig. 14 (which again comes from COCA).

By clicking on the name of a virtual corpus, they can delete, add, or move texts (from one virtual corpus to another, as shown in Fig. 13). They can see the number of texts and the size of the virtual corpus, as well as when the virtual corpus was created. They can delete the virtual corpus (delete symbol), and also "ignore" it (lock symbol),

TELESCOPE (RENAME) DELETE ADD TO MOVE TO --SELECT-- (SEE ALL VIRTUAL CORPORA)

HELP	☐ 100	YEAR	GENRE	SOURCE	TITLE
1	☐	1990	ACAD	Mercury	The Space Telescope: Eyes Above the Atmosphere....
2	☐	1990	MAG	ScienceNews	Dawn of a big telescope. (cover story)...
3	☐	1990	MAG	ScienceNews	Space Telescope: A Saga of Setbacks. (cover story)...
4	☐	1990	NEWS	NYTimes	New European Telescope Called Equal to Hubble...
5	☐	1990	SPOK	ABC_Nightline	The Hubble Telescope Launch
6	☐	1990	SPOK	PBS_Newshour	Newshour 900409
7	☐	1990	SPOK	PBS_Newshour	Newshour 900628
8	☐	1991	ACAD	Mercury	Discovering the invisible universe. (cover story)...
9	☐	1991	ACAD	Mercury	Starry, starry night: Observing on Kitt Peak

Fig. 13 Modifying virtual corpora

HELP		↕	↕	LIST NAME ↕	# TEXTS ↕	# WORDS ↕	FIND KEYWORDS ◉ SPECIFIC ○ FREQ	CREATED ↕
1			Sc	ASTRONOMY	327	249,396	NOUN VERB ADJ ADV N+N ADJ+N	1992 d
2				BELTANDROAD	100	121,103	NOUN VERB ADJ ADV N+N ADJ+N	748 d
3				BLM	100	345,984	NOUN VERB ADJ ADV N+N ADJ+N	125 d
4			Sc	CLIMATECHANGE	100	409,007	NOUN VERB ADJ ADV N+N ADJ+N	1742 d
5			Me	CORONAVIRUS	498	4,077,197	NOUN VERB ADJ ADV N+N ADJ+N	579 d
6			Me	COVID	400	3,620,882	NOUN VERB ADJ ADV N+N ADJ+N	578 d
7				COVID	1000	528,613	NOUN VERB ADJ ADV N+N ADJ+N	378 d
8				ELECTRON	125	87,460	NOUN VERB ADJ ADV N+N ADJ+N	2009 d
9			Po	IMMIGRANT	146	186,769	NOUN VERB ADJ ADV N+N ADJ+N	1440 d
10			Fi	INVESTMENT	98	40,037	NOUN VERB ADJ ADV N+N ADJ+N	2009 d

Fig. 14 Organising virtual corpora

so that it doesn't appear in the list of virtual corpora in main search form, and the virtual corpus will not be used when comparing the frequency of words in different virtual corpora (see Sect. 5.2). Via this page, users can also see the keywords from the virtual corpus (see Sects. 4.1 and 4.2 below). Users can also create a category for the virtual corpus (e.g. above Fi = Financial, Sc = Science, etc.), and they can then group their virtual corpora by clicking on the header for this column. For example, if they have 160 virtual corpora, they can group all of the ones dealing with medicine together, or (if they have specified this level of detail) all of the virtual corpora dealing with cardiology.

4 Keywords/Extracting Terms from the Virtual Corpora

4.1 Keyword Lists

For many users, the main purpose of creating virtual corpora is to extract specialised terminology related to their area of interest—e.g. aerospace engineering, immunology, international treaties, or investments. With the corpora from English-Corpora.org, this is both very fast and very easy.

To generate a keyword list, users simply click on NOUN, VERB, ADJ(ective), ADV(erb), N + N (noun + noun multiword expressions), or ADJ + N (adjective + noun) on the page shown in Fig. 14. In *less than a second*, they will see a list of keywords, as in Fig. 15 (nouns from the [nuclear power] virtual corpus shown in Fig. 2c). For iWeb (14 billion words from web pages), they see the number of tokens (FREQ) in their virtual corpus, the number of websites in the virtual corpus that have the word, the overall frequency of the word in the corpus, and the frequency compared to the "expected" frequency in their virtual corpus.

For example, if there are 14 million words in their virtual corpus, this is 0.1% of the entire 14 billion word corpus. Any word in their virtual corpus should have a frequency of 0.1% of the overall frequency in iWeb. If a word occurs 8,000 times in iWeb, then it should occur 8 times in the virtual corpus. If it occurs 12 times, then its frequency is 1.5× the expected frequency.

NUCLEAR_POWER [5,472,029 WORDS, 20 WEBSITES] (0.0% OF TOTAL) NOUN VERB ADJ ADV N+N ADJ+N [ALL CORPORA] SAVE LIST

HELP	ENTRY	WORD (CLICK FOR CONTEXT)	FREQ	# WEBSITES	SPECIFIC FREQ 90 10 WEBSITES	ENTIRE CORPUS	EXPECTED
1		NONPROLIFERATION	1816	16	1,343.6	3,458	1.4
2		RADIONUCLIDE	579	12	861.3	1,720	0.7
3		REACTOR	17385	18	848.7	52,409	20.5
4		RADIOISOTOPE	389	11	743.8	1,338	0.5
5		PLUTONIUM	3989	17	735.6	13,874	5.4
6		REPROCESSING	1337	13	573.5	5,965	2.3
7		HEXAFLUORIDE	169	10	514.1	841	0.3
8		WARHEAD	1809	11	510.5	9,066	3.5
9		DISARMAMENT	2162	10	462.1	11,970	4.7
10		URANIUM	9072	18	443.1	52,381	20.5
11		FISSION	1845	13	356.2	13,253	5.2

Fig. 15 Keyword lists: for [nuclear power] corpus in iWeb

Users can decide how "specific" of terms they want to extract. For example, the words in Fig. 15 are generated using the default setting. For a virtual corpus whose size is about 5–6 million words and 20 websites, the default would be a frequency of 90 tokens, and a range of 10 websites (from the total of 20 websites in the virtual corpus). But users could decide, for example, that they want to find words that are less frequent/more specialised, or else more frequent/less specialised. In Fig. 16, for example, they decide that the word only needs to occur 70 times, in just 4 of the 20 websites in the virtual corpus, and this will generate a more specialised list.

The ability to determine the level of specificity of the keywords of course applies to all corpora. For example, Fig. 17 is the default list from the 818,000 word virtual corpus from 100 texts in NOW (17.4 billion words total) from late 2015, dealing with [refugees]. The default settings are a token frequency of 90, and a range of 10 of the 100 texts.

Users could change this to just 25 tokens in 3 of the 100 texts, and this would produce a more specific keyword list (Fig. 18).

Or they could say that the word must appear at least 200 times in the 818,000 words, in 70 of the 100 texts. Because each of the words occurs in most of the 100

NUCLEAR_POWER [5,472,029 WORDS, 20 WEBSITES] (0.0% OF TOTAL) NOUN VERB ADJ ADV N+N ADJ+N [ALL CORPORA] SAVE LIST

HELP	ENTRY	WORD (CLICK FOR CONTEXT)	FREQ	# WEBSITES	SPECIFIC FREQ 70 4 WEBSITES	ENTIRE CORPUS	EXPECTED
1		TAILING	273	6	49,890.1	14	0.0
2		MILLISIEVERT	110	4	1,851.5	152	0.1
3		NONPROLIFERATION	1816	16	1,343.6	3,458	1.4
4		CONFINE	77	11	1,076.5	183	0.1
5		KILOTON	187	6	918.3	521	0.2
6		RADIONUCLIDE	579	12	861.3	1,720	0.7
7		REACTOR	17385	18	848.7	52,409	20.5
8		RADIOISOTOPE	389	11	743.8	1,338	0.5
9		PLUTONIUM	3989	17	735.6	13,874	5.4
10		REPROCESSING	1337	13	573.5	5,965	2.3
11		HEXAFLUORIDE	169	10	514.1	841	0.3

Fig. 16 More specific keyword lists: [nuclear power] in iWeb

REFUGEES-2015 [818,631 WORDS, 100 TEXTS] (6.0% OF TOTAL) NOUN VERB ADJ ADV N+N ADJ+N [ALL CORPORA] SAVE LIST

HELP	WORD (CLICK FOR CONTEXT)	FREQ	# TEXTS	SPECIFIC FREQ 90 10 TEXTS	ENTIRE CORPUS	EXPECTED
1	REFUGEE	6523	96	0.7	150,215	9,042.0
2	ASYLUM	1612	68	0.7	41,136	2,476.1
3	RESETTLEMENT	281	46	0.6	7,598	457.3
4	SEEKER	908	53	0.5	30,492	1,835.4
5	SMUGGLER	217	28	0.3	13,659	822.2
6	FEE	3015	10	0.2	227,624	13,701.5
7	PERSECUTION	198	43	0.2	15,778	949.7
8	DISPLACEMENT	106	27	0.2	9,888	595.2
9	MIGRANT	846	62	0.2	80,739	4,860.0
10	APPLICANT	574	20	0.2	56,254	3,386.1
11	MIGRATION	371	43	0.1	44,254	2,663.8

Fig. 17 Keyword lists: for [refugees] corpus in NOW

REFUGEES-2015 [818,631 WORDS, 100 TEXTS] (6.0% OF TOTAL) NOUN VERB ADJ ADV N+N ADJ+N [ALL CORPORA] SAVE LIST

HELP	WORD (CLICK FOR CONTEXT)	FREQ	# TEXTS	SPECIFIC FREQ 25 3 TEXTS	ENTIRE CORPUS	EXPECTED
1	NONIMMIGRANTS	32	3	33.2	16	1.0
2	NON-REFOULEMENT	43	11	8.6	83	5.0
3	NATURALIZATION	301	3	4.3	1,169	70.4
4	ADJUDICATION	217	4	1.1	3,336	200.8
5	REFUGEE	6523	96	0.7	150,215	9,042.0
6	ASYLUM	1612	68	0.7	41,136	2,476.1
7	RESETTLEMENT	281	46	0.6	7,598	457.3

Fig. 18 More specific keyword lists: [refugees] in NOW

texts (at least 70 texts), these will be high frequency, commonly-occurring words in the language (Fig. 19).

As a final example, the keyword list in Fig. 20 comes from the 221,500 word virtual corpus dealing with [endocrinology] in 100 texts in the two billion word Wikipedia corpus. The default values are 45 tokens in 10 of the 100 texts.

Figure 21 shows the words that occur less times in a smaller number of texts (just 30 tokens in the 221,500 word virtual corpus, in just 3 of the 100 texts), and of course the word list is even more specific to endocrinology.

REFUGEES-2015 [818,631 WORDS, 100 TEXTS] (6.0% OF TOTAL) NOUN VERB ADJ ADV N+N ADJ+N [ALL CORPORA] SAVE LIST

HELP	WORD (CLICK FOR CONTEXT)	FREQ	# TEXTS	SPECIFIC FREQ 200 70 TEXTS	ENTIRE CORPUS	EXPECTED
1	REFUGEE	6523	96	0.7	150,215	9,042.0
2	CAMP	701	72	0.1	205,479	12,368.5
3	COUNTRY	2207	96	0.0	2,128,105	128,098.0
4	GOVERNMENT	1832	86	0.0	2,376,887	143,073.0
5	NUMBER	985	84	0.0	1,511,895	91,006.2
6	WAR	487	70	0.0	748,700	45,066.8
7	PEOPLE	2976	89	0.0	4,745,533	285,650.0
8	ISSUE	575	73	0.0	1,094,490	65,881.1
9	MILLION	764	81	0.0	1,525,671	91,835.4
10	FAMILY	754	70	0.0	1,652,429	99,465.4

Fig. 19 More general keyword lists: [refugees] in NOW

ENDOCRIN_ [221,533 WORDS, 100 TEXTS] NOUN VERB ADJ ADV N+N ADJ+N [ALL CORPORA] SAVE LIST

HELP	WORD (CLICK FOR CONTEXT)	FREQ	# TEXTS	SPECIFIC FREQ 45 10 TEXTS	ALL WIKIPEDIA	EXPECTED
1	NEOPLASM	318	10	4,017.5	661	0.1
2	DISRUPTOR	68	14	1,426.8	398	0.0
3	NEOPLASIA	66	14	1,371.0	402	0.0
4	ENDOCRINE	46	24	1,231.2	312	0.0
5	ENDOCRINOLOGIST	51	22	1,179.8	361	0.0
6	ADENOMA	68	14	1,128.9	503	0.1
7	PITUITARY	73	18	1,033.2	590	0.1
8	HORMONE	763	71	726.0	8,776	1.1
9	ANDROGEN	100	19	651.9	1,281	0.2
10	PANCREAS	159	21	594.6	2,233	0.3

Fig. 20 Keyword lists: for [endocrinology] corpus in Wikipedia

ENDOCRIN_ [221,533 WORDS, 100 TEXTS] NOUN VERB ADJ ADV N+N ADJ+N [ALL CORPORA] SAVE LIST

HELP	WORD (CLICK FOR CONTEXT)	FREQ	# TEXTS	SPECIFIC FREQ 30 3 TEXTS	ALL WIKIPEDIA	EXPECTED
1	NEOPLASM	318	10	4,017.5	661	0.1
2	XENOESTROGENS	31	7	3,595.5	72	0.0
3	ENDOSULFAN	56	3	2,718.9	172	0.0
4	PHEOCHROMOCYTOMA	35	8	1,614.8	181	0.0
5	BISPHENOL	56	5	1,452.3	322	0.0
6	DISRUPTOR	68	14	1,426.8	398	0.0
7	NEOPLASIA	66	14	1,371.0	402	0.0
8	NEVUS	99	3	1,291.8	640	0.1
9	TESTE	37	15	1,271.5	243	0.0

Fig. 21 More specific keyword lists: [endocrinology] in NOW

4.2 Multiword Expressions

In addition to extracting single word terms, it is also possible to extract multiword expressions (MWEs)—both NOUN + NOUN and ADJ-NOUN. For example, Fig. 22a shows the NOUN + NOUN MWEs from a 50 million word virtual corpus from 10 websites in iWeb, dealing with [investments] (especially investment in mining companies); Fig. 22b is NOUN + NOUN for [linguistics] in Wikipedia; Fig. 22c is ADJ + NOUN for [hormones] in iWeb, and Fig. 22d is ADJ + NOUN for [ventilators] in the Coronavirus Corpus. As with the single word keyword lists, users can also adjust the specificity of the MWEs, using the number of tokens and the number of texts (or websites, in iWeb).

It may be worth mentioning that all of these keyword lists (both single words and multiword expressions) are extracted almost instantaneously from the corpora, even when the corpus itself is billions of words in size, and the virtual corpus is 30–50 million words or larger. Again, this is due to the underlying architecture for the corpora. The architecture relies on relational databases, and part of the information that they store is the frequency of each noun, verb, adjective, adverb, noun + noun, and adj + noun in each text in the corpus—even millions of texts. Once we have created a virtual corpus (which usually takes less than one second), then it is just

a

INVESTMENT2021 [49,923,337 WORDS, 10 WEBSITES] (0.4% OF TOTAL) NOUN VERB ADJ ADV N+N ADJ+N [ALL CORPORA] SAVE LIST

HELP	ENTRY	WORD (CLICK FOR CONTEXT)	FREQ	# WEBSITES	SPECIFIC FREQ 30 7 WEBSITES	ENTIRE CORPUS	EXPECTED
1		COMPANY PROFILE	9785	8	144.4	19,003	67.8
2		SECURITY LEGISLATION	1227	7	135.9	2,532	9.0
3		SHARE PURCHASE	1449	8	119.8	3,391	12.1
4		MINERAL RESOURCE	3705	9	101.4	10,242	36.5
5		INDUSTRY CONDITION	369	7	100.3	1,032	3.7
6		SHAREHOLDER APPROVAL	1081	7	95.1	3,189	11.4
7		GOLD PROJECT	1425	7	94.9	4,213	15.0
8		MINERAL EXPLORATION	1070	7	92.3	3,252	11.6
9		SECURITY REGULATOR	843	8	90.4	2,616	9.3
10		SECURITY LAW	5073	9	86.0	16,549	59.0

b

LINGUISTICS [139,957 WORDS, 98 TEXTS] NOUN VERB ADJ ADV N+N ADJ+N [ALL CORPORA] SAVE LIST

HELP	WORD (CLICK FOR CONTEXT)	FREQ	# TEXTS	SPECIFIC FREQ 6 3 TEXTS	ALL WIKIPEDIA	EXPECTED
1	LANGUAGE ACQUISITION	16	8	52,873.4	4	0.0
2	LANGUAGE FAMILY	28	10	52,873.4	7	0.0
3	SPEECH COMMUNITY	19	9	50,229.7	5	0.0
4	STRUCTURE GRAMMAR	14	10	30,842.8	6	0.0
5	WORD COMBINATION	7	4	30,842.8	3	0.0
6	SOUND CHANGE	7	4	30,842.8	3	0.0
7	LANGUAGE USE	28	12	23,132.1	16	0.0
8	LANGUAGE USER	6	6	19,827.5	4	0.0
9	LANGUAGE POLICY	6	4	15,862.0	5	0.0
10	WORD FORMATION	25	5	14,367.8	23	0.0

c

HORMONE [6,724,164 WORDS, 10 WEBSITES] (0.0% OF TOTAL) NOUN VERB ADJ ADV N+N ADJ+N [ALL CORPORA] SAVE LIST

HELP	ENTRY	WORD (CLICK FOR CONTEXT)	FREQ	# WEBSITES	SPECIFIC FREQ 30 6 WEBSITES	ENTIRE CORPUS	EXPECTED
1		VAGINAL DRYNESS	457	6	267.7	3,554	1.7
2		HOT FLASH	1426	9	239.3	12,407	6.0
3		MENOPAUSAL WOMAN	285	6	229.8	2,582	1.2
4		EARLY MENOPAUSE	162	6	198.5	1,699	0.8
5		UNDERACTIVE THYROID	167	7	194.4	1,789	0.9
6		RADIOACTIVE IODINE	295	6	178.0	3,450	1.7
7		NATURAL HORMONE	173	7	171.2	2,104	1.0
8		MENOPAUSAL SYMPTOM	256	7	158.6	3,361	1.6
9		HORMONAL IMBALANCE	568	7	155.2	7,622	3.7
10		OVERACTIVE THYROID	91	7	151.3	1,252	0.6
11		ADRENAL INSUFFICIENCY	138	6	151.0	1,903	0.9
12		ENDOCRINE GLAND	166	6	147.3	2,347	1.1

d

VENTILATOR [470,541 WORDS, 100 TEXTS] (0.0% OF TOTAL) NOUN VERB ADJ ADV N+N ADJ+N [ALL CORPORA] SAVE LIST

HELP	WORD (CLICK FOR CONTEXT)	FREQ	# TEXTS	SPECIFIC FREQ 6 10 TEXTS	ENTIRE CORPUS	EXPECTED
1	MEDICAL DEVICE	49	21	21,243.6	6	0.0
2	SEVERE CASE	16	14	10,405.0	4	0.0
3	MEDICAL EQUIPMENT	34	17	9,827.0	9	0.0
4	RESPIRATORY DISTRESS	18	13	6,689.0	7	0.0
5	INTENSIVE CARE	94	38	4,289.8	57	0.0
6	CRITICAL CARE	32	14	1,632.2	51	0.0
7	FEDERAL GOVERNMENT	178	17	1,129.3	410	0.2
8	RESPIRATORY FAILURE	16	12	513.8	81	0.0
9	INFECTIOUS DISEASE	21	10	510.5	107	0.0
10	DAILY BRIEFING	10	10	490.8	53	0.0
11	MEDICAL SUPPLY	31	12	438.3	184	0.1
12	FRONT LINE	17	11	255.6	173	0.1

Fig. 22 Multiword lists **a** NOUN + NOUN for [investments/mining] in COCA **b** NOUN + NOUN for [linguistics] in Wikipedia **c** ADJ + NOUN for [hormones] in iWeb **d** ADJ + NOUN for [ventilators] in the coronavirus corpus

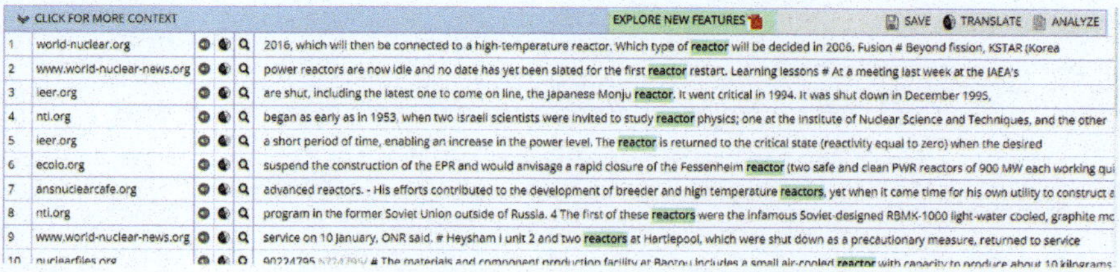

Fig. 23 KWIC display for reactor in the [nuclear power] virtual corpus in iWeb

another second or so to compare the words and phrases in that virtual corpus to the overall corpus, and to display these in the keyword lists.

4.3 *Word and Phrase-Based Resources*

The lists of keywords and multiword expressions are linked to extremely rich resources for each word and phrase. From the list of keywords, users can click on a word to see it in context in the virtual corpus, as with *reactor* in Fig. 23.

From the Keyword in Context display, they can also hear any of the entries pronounced, get a translation to one of 120 different languages, save entries for later analysis, and create an even more specific "virtual corpus" from the "virtual corpus" (e.g. entries for *reactor* in the [nuclear power] corpus), and then analyse that very specialised corpus.

With one click, they can also get detailed information on any of the keywords from the virtual corpus, as with *reactor* (Fig. 24).

This "word sketch" contains definitions, links to images, videos, and pronunciation; synonyms; semantically-related topics and morphologically-related words; word clusters and concordance lines; and collocates. This "word page" is just a summary of even more detailed pages, such as the collocates page for *reactor*, as shown in Fig. 25.

5 Searching Within and Comparing Virtual Corpora

5.1 In addition to creating lists of terms from the virtual corpora, users can also search within a virtual corpus. For example, Fig. 26 shows the most frequent strings for the search *ADJ object* (adjective + *object*) in all of COCA.

But it is possible to search just within a virtual corpus. For example, Fig. 27 shows the most frequent strings for *ADJ object* in a virtual corpus composed of 1,852,000 words in 879 texts, which was created by searching for texts from the magazine *Astronomy*. Note that there is no overlap between the 11 most frequent strings of *ADJ object* in the [astronomy] virtual corpus and the 11 most frequent strings in all

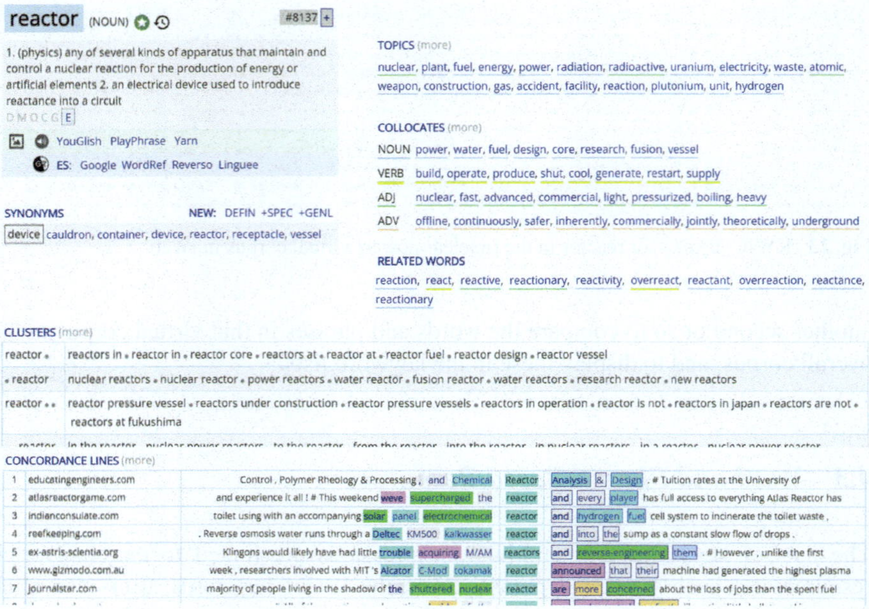

Fig. 24 "Word sketch" for *reactor* in iWeb

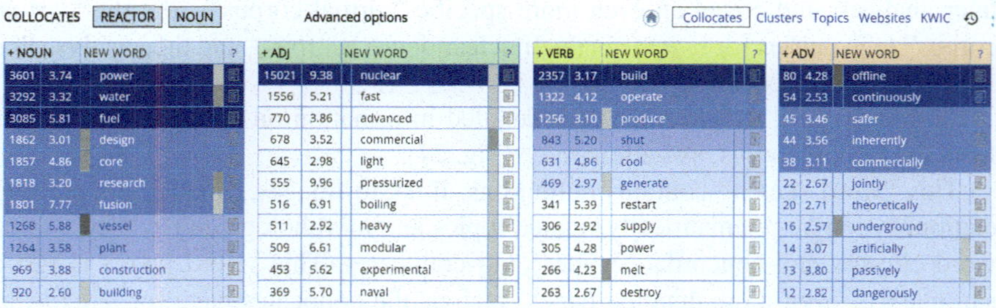

Fig. 25 Collocates of reactor in iWeb

of COCA. This shows the value of limiting searches to a particular topic, where the language can be quite different than in a "general purpose" corpus like COCA.

And it is not just words and phrases that are different in a virtual corpus for a particular topic. There are also semantic differences, as measured via collocates. For example, in Wikipedia, we can create two virtual corpora, one dealing with [engineering] and the other with [psychology]. When we search for collocates of *stress* in these two virtual corpora, we see that there is virtually no overlap in the meaning of *stress*, at least as measured by collocates (Fig. 28).

5.2 Because we can easily find the frequency of words and phrases in different virtual corpora, it is also possible to compare the frequency of different terms in the several corpora that we have created. For example, we could have a list of 15–20

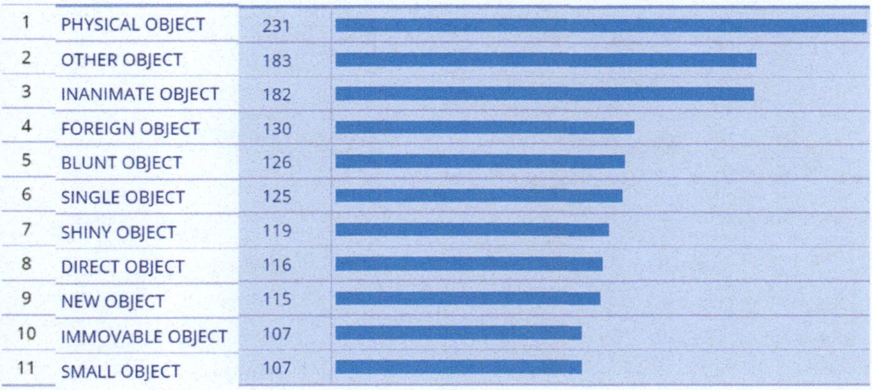

Fig. 26 ADJ + object in COCA: entire corpus

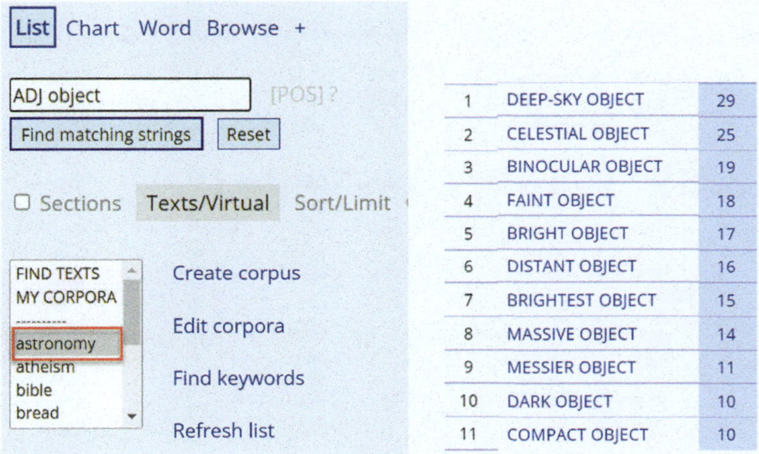

Fig. 27 ADJ + object in COCA: [astronomy] virtual corpus

engineering terms, and then easily see their frequency in Wikipedia in four virtual corpora related to engineering—civil engineering, chemical engineering, mechanical engineering, and electrical engineering, to tease apart the lexical differences between these related fields. As Fig. 29a–d show, we would find that *equilibrium* is the most common in electrical engineering, *resonance* in mechanical engineering, *gradient* in civil engineering, and *measurement* in electrical engineering.

It is not that each of these words doesn't occur in the other engineering virtual corpus (they do), but each takes on a particular importance and salience in a particular field of engineering.

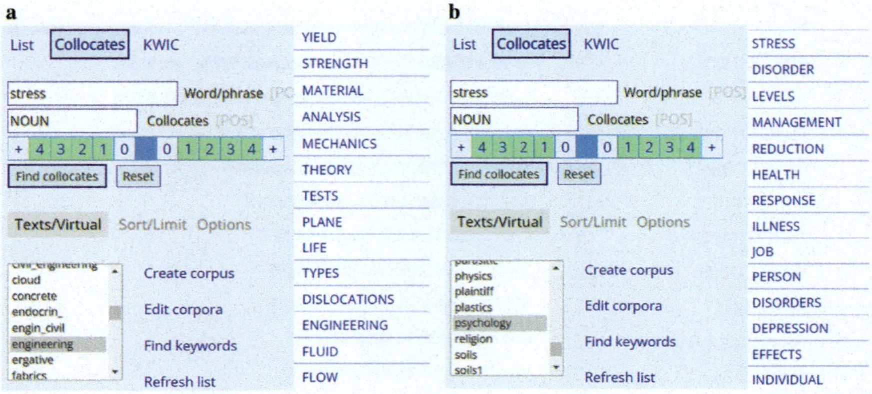

Fig. 28 Collocates of stress in Wikipedia **a** engineering **b** psychology

a

HELP	100	MORE SEARCHES	SEE WORD IN CONTEXT	SEE PAGE	# WORDS	# HITS	RELEVANCE	PER MILLION WORDS
1			ENG-CHEM		103,918	157	1,510.8	
2			ENG-MECH		37,486	20	533.5	
3			ENG-CIVIL		152,471	24	157.4	
4			ENG-ELEC		131,453	2	15.2	

b

HELP	100	MORE SEARCHES	SEE WORD IN CONTEXT	SEE PAGE	# WORDS	# HITS	RELEVANCE	PER MILLION WORDS
1			ENG-MECH		37,486	48	1,280.5	
2			ENG-ELEC		131,453	6	45.6	
3			ENG-CIVIL		152,471	3	19.7	
4			ENG-CHEM		103,918	1	9.6	

c

HELP	100	MORE SEARCHES	SEE WORD IN CONTEXT	SEE PAGE	# WORDS	# HITS	RELEVANCE	PER MILLION WORDS
1			ENG-CIVIL		152,471	26	170.5	
2			ENG-CHEM		103,918	10	96.2	
3			ENG-MECH		37,486	3	80.0	
4			ENG-ELEC		131,453	8	60.9	

d

HELP	100	MORE SEARCHES	SEE WORD IN CONTEXT	SEE PAGE	# WORDS	# HITS	RELEVANCE	PER MILLION WORDS
1			ENG-ELEC		131,453	76	578.2	
2			ENG-CIVIL		152,471	32	209.9	
3			ENG-MECH		37,486	4	106.7	
4			ENG-CHEM		103,918	9	86.6	

Fig. 29 Frequency of words in [chemical], [electrical], [mechanical], and [civil] engineering virtual corpora in Wikipedia **a** equilibrium **b** resonance **c** gradient **d** measurement

6 Conclusion

Previous researchers have shown the value of creating corpora of specific domains (e.g. biology, medicine, engineering, or finance) and then creating word lists and lists of multiword expressions from these corpora. Smith (2020) has described how this can be done efficiently using Sketch Engine and BootCat.

In this chapter, we have shown that it is both very easy and very quick to create "virtual corpora" in the corpora from English-Corpora.org. In just a few seconds and with just a few clicks, users can create corpora on almost any topic. With more than 22.3 million texts in iWeb, 24.1 million texts in NOW (including 3.8 million from the last year), and millions of texts from the other 15 English corpora, there is rich data on virtually any topic. And in just one or two seconds more, they can then extract key terms from these corpora, including multiword expressions. And of course, they can easily search within these virtual corpora (to find phrases and collocates), and even compare word and phrase frequency between different virtual corpora.

Corpus linguists usually like to create corpora, and we assume that others want to do the same—even if it takes an hour or a day or more, and even if we need to learn to use special software to do so. But translators and teachers and students of English for Specific Purposes are not corpus linguists, and most of them just want to create high quality corpora as quickly and as easily as possible.

With English-Corpora.org, these researchers can search through tens of billions of words in tens of millions of texts, to create these specialised corpora much quicker and much more easily than with any other tool. The end result is that they are then able to focus on the tasks that they really do care about—the extraction and use of keywords and phrases for particular domains.

References

Ackermann, Kirsten, and Yu-Hua Chen. 2013. Developing the academic collocation list (ACL): A corpus-driven and expert-judged approach. *Journal of English for Academic Purposes* 12: 235–247.

Baroni, Marco, and Silvia Bernardini. 2004. BootCaT: Bootstrapping corpora and terms from the web. In *Proceedings of 4th International Conference on Language Resources and Evaluation*, 1313–1316. Lisbon, Portugal.

Baroni, Marco, Adam Kilgarriff, Jan Pomikálek, and Pavel Rychlý. 2006. WebBootCaT: Instant domain-specific corpora to support human translators. In *Proceedings, 11th Annual Conference of the European Association for Machine Translation Conference*, 247–252. Oslo, Norway.

Browne, Charles, Brent Culligan, and Joseph Phillips. 2013. *A New Academic Word List*. http://www.newgeneralservicelist.org/nawl-newacademic-word-list. Accessed October 27, 2021.

Castagnoli, Sara. 2006. Using the web as a source of LSP corpora in the terminology classroom. In *Wacky! Working Papers on the Web as Corpus*, ed. Marco Baroni, and Silvia Bernardini, 159–172. Bologna: Gedit.

Charles, Maggie. 2014. Getting the corpus habit: EAP students' long-term use of personal corpora. *English for Specific Purposes* 35: 30–40.

Coxhead, Averil. 2000. A new academic word list. *TESOL Quarterly* 34: 213–238.

Davies, Mark. 2008a. English-Corpora.org. Accessed October 27, 2021.

Davies, Mark. 2008b. *The Corpus of Contemporary American English (COCA)*. Available online at https://www.english-corpora.org/coca/

Davies, Mark. 2019. *Word Frequency Data from the Corpus of Contemporary American English*. https://www.wordfrequency.info. Accessed October 27, 2021.

Davies, Mark. 2017. Using large online corpora to examine lexical, semantic, and cultural variation in different dialects and time periods. In *Corpus-Based Sociolinguistics*, ed. Eric Friginal et al., 19–82. London: Routledge.

Davies, Mark. 2018. Corpus-based studies of lexical and semantic variation: The importance of both corpus size and corpus design. In *From Data to Evidence in English Language Research* (Digital Linguistics), ed. Suhr, Carla, Terttu Nevalainen, and Irma Taavitsainen, 34–55. Leiden: Brill.

Gardner, Dee, and Mark Davies. 2013. A new academic vocabulary list. *Applied Linguistics* 35: 1–24.

Kilgarrif, Adam, Pavel Rychlý, Pavel Smrz, and David Tugwell. 2004. The sketch engine. In *EURALEX 2004 Proceedings, Lorient, France.*

Lee, David, and John Swales. 2006. A corpus-based EAP course for NNS doctoral students: Moving from available specialised corpora to self-compiled corpora. *English for Specific Purposes* 25: 56–75.

Mudraya, Olga. 2006. Engineering English: A lexical frequency instructional model. *English for Specific Purposes* 25: 235–256.

Simpson-Vlach, Rita, and Nick C. Ellis. 2010. An academic formulas list: New methods in phraseology research. *Applied Linguistics* 31: 487–512.

Smith, Simon. 2015. Construction and use of thematic corpora by academic English learners. In *Task Design & CALL: Proceedings of 17th International CALL Conference*, ed. Jozef Colpaert, Ann Aerts, Margret Oberhofer, and Mar Gutiérez-Colón Plana, 437–445. Tarragona, Antwerp: Universiteit Antwerpen.

Smith, Simon. 2020. DIY corpora for accounting & finance vocabulary learning. *English for Specific Purposes* 57: 1–12.

West, Michael. 1953. *A General Service List of English Words*. London: Longman.

Mark Davies is a Professor Emeritus of Linguistics at Brigham Young University in Provo, Utah, USA. He has published widely on language variation and change, and he has received several large grants to create and analyse corpora. He is the creator of several large corpora and corpus-based tools that are available from English-Corpora.org, which are used by hundreds of thousands of researchers, teachers, and translators each month.

Working with Corpora in Translation Technology Teaching: Enhancing Aspects of Course Design

Mark Shuttleworth

1 Introduction

This article is derived from teaching materials that were originally created for a translation technology course at UCL and are currently being used for the same purpose at Hong Kong Baptist University. It aims to explore two topics that involve the use of corpora—in terms of how they can be implemented and of their potential place in a translation technology curriculum. The first of these topics—which is considered in Sects. 2 and 3—is term extraction, and the second is the acquisition of bilingual data. In the case of the first of these, the topic of personality typology—in the guise of the Myers-Briggs Type Indicator (MBTI)—has been chosen for illustration purposes, partly because of the intrinsic interest it is likely to hold for many participants, but perhaps more importantly because it is not only rich in terminology but also conceptually relatively accessible. Consequently, these sections are largely illustrated by the use of an HTML document that was constructed from a number of short texts downloaded from the internet in 2018 (from a source that is no longer available) and comprising 13,700 words of text discussing the sixteen different personality types that are identified by the MBTI. These sections consider the use of both a CAT tool and a dedicated suite of lexical analysis tools and how they might be deployed within a translation technology course. When a language other than English is included, it is usually Chinese.

In Sect. 4, the focus shifts to the second topic, that of locating and acquiring bilingual data, for example, for the purpose of enhancing the performance of CAT tools. Regarding this topic, it was Pierre Isabelle and his associates who stated in 1993 that 'existing translations contain more solutions to more translation problems than any other available resource' (Isabelle et al. 1993: 205). Indeed, as is well known, parallel text is one of the main engines that drives the use and advance of translation

M. Shuttleworth (✉)
Hong Kong Baptist University, Kowloon, Hong Kong, China
e-mail: markshut@hkbu.edu.hk

J. Pan and S. Laviosa (eds.), *Corpora and Translation Education*, New Frontiers in Translation Studies, https://doi.org/10.1007/978-981-99-6589-2_6

technology. This section aims to work out some of the practical implications of this quotation by investigating the potential firstly of a single highly significant data repository and secondly of other alternative online resources for exploitation within a translation technology course, and also how we introduce our students to the idea of working with (relatively) big data.

Since translation technology started to be taught as a subject at university in the middle to late 1990s, it has developed to such an extent that in many countries, it is seen as an indispensable part of university-level translator training (Chan and Shuttleworth 2023). Within many such courses, it is a technology known as translation memory (TM) that frequently occupies the most important place (Rothwell and Svoboda 2019; Zhang and Vieira 2021). Translation memory generally forms the centre-piece of computer-assisted translation (CAT) tools, while a TM represents one of the clearest embodiments of what is envisaged by the words of Isabelle et al. TM remembers a translator's past translation solutions and suggests them back when similar wording is encountered on a subsequent occasion. In addition, most CAT tools permit you to search the TM interactively via a concordance function. The TM concept becomes more powerful when TMs are shared, either in real time across members of a team or else between tools, individuals or organisations, usually by means of the XML-based TMX (Translation Memory eXchange) file format.

Terminology management forms a standard component in many translation technology courses (Rodríguez-Castro 2018: 360–1; Rothwell and Svoboda 2019: 52), not least because this activity generally forms an indispensable part of most types of specialised translation. While professional terminologists will typically work with a particular technical terminology in a more systematic manner, the terminological work that most translators engage in is more ad hoc in nature as it aims to solve problems that arise within a particular translation job (Wright and Budin 1996: 150). One way to do this is via a process known as terminology (or term) extraction, in which software solutions are applied in order to identify candidate terms (Korkontzelos and Ananiadou 2022). In line with the nature of terminology, these may be either single or multi-word items and are almost always nouns rather than belonging to a different part of speech. Term extraction works according to either linguistic, dictionary-based or statistical principles. The systems described in Sects. 2 and 3 use the third approach, which, because no language-specific information is needed, can work equally well for any language combination—in line with most CAT tools. In addition, these systems work on the basis of wordlists and keyword lists respectively, the advantage of the latter approach being that it allows lexical items that are somehow characteristic of a corpus to be identified with ease because of their relative "keyness" (Bondi and Scott 2010), the assumption being that the majority of terms are likely to be marked in this way. Once the term extraction process has been conducted, however, it will always be necessary for the list of candidate terms that is produced to be checked by a human, who will make decisions as to which should be retained for further processing and which rejected. Systems for monolingual term extraction are relatively widely available, while those capable of performing the more complex process of bilingual term extraction are relatively few in number.

There is evidence that term extraction also occupies a significant place in many course curricula; the 2017 European Master's in Translation survey, for example, reveals that, at that time, it was a compulsory feature of 67% of programmes (Rothwell and Svoboda 2019: 37). In the light of this, the present article seeks to document what are likely to be some of the more frequently opted-for tools and approaches for covering this part of the curriculum and to reflect on their potential and relative effectiveness.

Moving on from this, one insight that has not perhaps been quite so widely implemented across translation technology courses is that it is possible to grow students' TMs way beyond the starter level and thus equip them with resources that can help them to appreciate the full potential of the technology. Relatively little has been written about the place of parallel data acquisition in translation curricula, although there is a strong likelihood that teachers employ a number of methods to ensure the greater availability of existing translation assets for their students (Chan and Shuttleworth 2023: 271); these include the downloading of substantially-sized publicly available sets of parallel text (ibid.), although it is not believed that this approach is currently widely implemented in courses. The present article aims to begin to rectify this lack of theoretical discussion by taking a detailed look at a number of sites that offer large amounts of parallel data and also to consider some of the uses to which this may be put in order to improve students' learning experience by enabling them to make full use of software functions that might not otherwise work in a convincing manner. These would, in particular, include a CAT tool's concordance search, the feature that is known in Déjà Vu as Assemble from Portions, in memoQ as Fragment Assembly and in Trados Studio as Fragment Matches, and the Trados Studio AutoSuggest Dictionary, which can only be created from a TM that contains at least 10,000 translation units (RWS Documentation Center 2021). Needless to say, the availability of large TMs also potentially impacts a CAT tool's ability to provide useful fuzzy matches. Besides this, assuming the topic is relevant, a large parallel corpus can represent a highly suitable resource for systematic terminology work. And finally, for the purposes of MT engine training—a realistic proposition with the advent of systems such as KantanMT—resources that are used for training purposes also need to contain at least 10,000 segments.

2 Term Extraction with Phrase[1]

As is fairly well known, translators spend a considerable amount of time researching terminology, according to one writer ranging from 20 to 25% for established professionals to up to 40–60% for those who are less experienced (Champagne 2004: 30). Some translators will respond to this reality by investigating possible technological solutions of the type discussed below. Given its potential importance for professional translators and the opportunity it gives to provide our students with a clear

[1] This tool was known as Memsource until 2021.

time-saving benefit when they start working, the arguments for including this as part of a course in translation technology are strong.

It is probably the case that most typical CAT tools offer no term extraction facility, although there are a number of prominent exceptions. The obvious one is Multiterm Extract, a professional-level utility that offers both mono- and bilingual term extraction. XTM Cloud also offers bilingual functionality. This is likely to be the preferred tool of those who already use Trados Studio, although it has to be purchased separately at significant additional expense. Besides this, Déjà Vu and memoQ also offer monolingual extraction along very similar lines; this function is termed the "lexicon" in the former and "extract candidates" in the latter. In addition, Phrase also provides some monolingual term extraction functionality. By and large, however, if a user wants to be able to carry out bilingual term extract, he or she will either need to use Multiterm Extract or turn to an alternative, dedicated application.

Phrase has been selected as the focus of this section for two reasons. Firstly, it not only features prominently in our translation technology course at Hong Kong Baptist University but appears to be a popular choice for inclusion in curricula in a number of countries. Secondly, because the analysis it performs is based on wordlists rather than keyword lists, the way the extraction facility works forms a useful contrast with Sketch Engine,[2] which forms the focus of the next section. Phrase itself has been in existence since 2010 (Nimdzi Insights 2021) and is available both in the cloud and as a locally installed application. Since its launch, it has become one of the most popular CAT tools available, with over 250,000 users around the world as of 2021 (Nimdzi Insights 2021) including corporate customers such as Uber, Zendesk, Lionbridge and Huawei (Phrase 2023). The quoted figure compares favourably to Trados Studio, the company back in 2014 claiming to have sold 200,000 licences (including for Trados, SDLX and SDL Trados Studio, half of this number being for SDL Trados Studio) (Ghislandi 2014).

Term extraction in Phrase will normally be based on one or more source texts within the active project and will be of the "ad hoc" type that is normally pursued by translators. Although simple, the interface offers a number of options, as illustrated in Fig. 1.

The settings configured in Fig. 1 specify that candidate terms identified should be no more than three words in length, should occur no fewer than two times in the corpus being analysed and should contain no fewer than four letters. This effectively means that both single- and multi-word items can be identified, while rare items and many grammatical words (probably the greater number of words with fewer than four letters) will be excluded.

Carrying out an extraction with the above settings on the personality type corpus produces a wordlist containing both single- and multi-word term candidates. The results are saved in XSLX format and have to be viewed in Excel.

Because this is a wordlist rather than a list of keywords, nothing has been excluded beyond what was specified in the settings; in other words, salience or termhood are not criteria for inclusion. Indeed, some of the items listed in Fig. 2 are clearly not terms

[2] https://www.sketchengine.eu/.

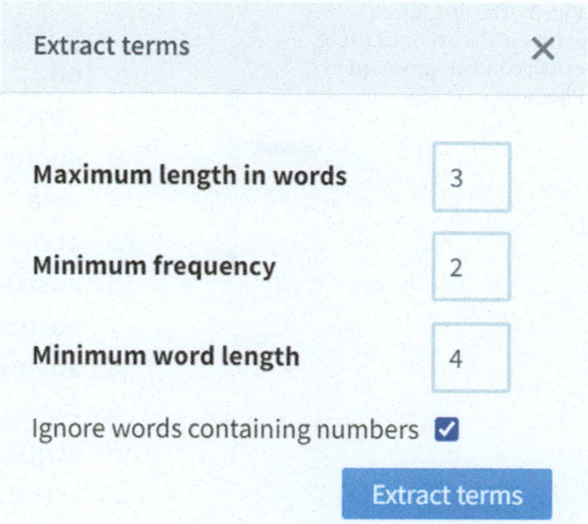

Fig. 1 Term extraction settings in Phrase

(e.g. *abilities*, *able* and *across*). Others are formal words or collocations (e.g. *abstract information* and *acronym*), but their usage is too general for them to be considered as terms. Finally, there are only likely to be a few that one would wish to confirm as terms (probably only *action-oriented* from Fig. 2). The reason for this is that no account has been taken of the keyness of the items, with the result that actual terms—the use of which is particularly characteristic of a specific text or subject-area—are not foregrounded in any way and we end up with an undifferentiated mix of terms and common lexical (and some grammatical) items.

This is as far as the term extraction process offered by Phrase will take you. A possible extension might be to permit bilingual extraction from a TM, although in developmental terms, this would be likely to require a significant amount of effort. As this is just a monolingual extraction, typical processing would then simply entail deleting a large proportion of the list so that only those judged to have terminological status remained. Following that, the resulting list of candidate terms would be used as the basis for independently researching the target language equivalents and entering them to produce a bilingual list. This could involve a range of options. For example, the free online utility Linguee.com permits a user to search a very large online multilingual corpus to obtain suggestions for target language equivalents. Along similar lines, but with greater flexibility, the Sketch Engine parallel concordance tool can be used in the same way but searching a corpus that you specify or one that you have constructed yourself. Entire specialised glossaries or dictionaries can be located using the lists supplied by sites such as Lexicool.com, while resources such as Babelnet.org provide multilingual equivalents as well as other secondary resources. For terms that have a Wikipedia article written on them, equivalents (and often considerable additional information) can be found by clicking on the interlanguage link for the relevant language, if there is one. Finally, Google or another search engine

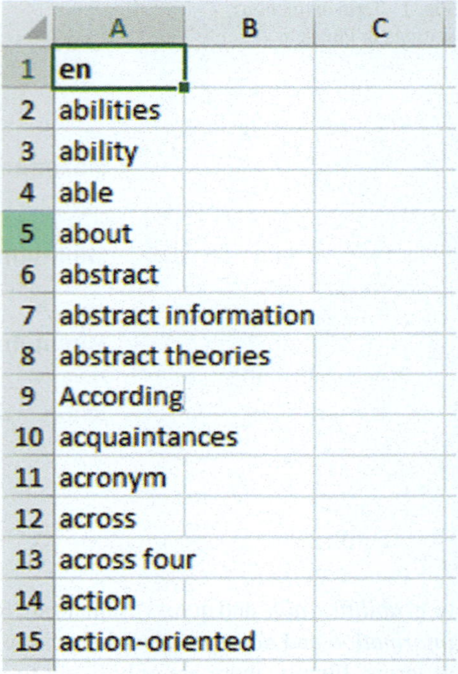

	A	B	C
1	en		
2	abilities		
3	ability		
4	able		
5	about		
6	abstract		
7	abstract information		
8	abstract theories		
9	According		
10	acquaintances		
11	acronym		
12	across		
13	across four		
14	action		
15	action-oriented		

Fig. 2 The first fifteen entries in the exported list of extracted terms produced by Phrase

can often be coaxed into providing a term equivalent that cannot be located in any other way.

The next section proposes an alternative to using a CAT tool—and one that can serve as a convenient entry-point to working with corpora.

3 Term Extraction with Sketch Engine

There is indeed a very effective alternative to using a CAT tool to extract terms, and that is utilising a corpus-analysis tool. Indeed, the explicit use of corpora is often identified as an important component of translation technology training. Rothwell and Svoboda, for example, reporting on the findings of the survey run by the European Master's in Translation in 2017, identify the use of corpus-analysis software as a major new teaching emphasis (2019: 40–41, 53). The online application Sketch Engine, which consists of a suite of tools designed for lexical analysis, provides access to corpora in more than sixty languages that are tagged with grammatical information. As will be seen, it can be used to produce much higher-quality lists of candidate terms as it employs an approach based on keywords. In addition, it has a bilingual term extraction capability. Besides this, it offers users the ability to create their own corpora, which permits them, for example, to carry out term extraction in a specific subject area.

3.1 Monolingual Term Extraction

As with Phrase above, Sketch Engine allows users to process texts for the purposes of term extraction. However, the differences between these two systems are considerable. Phrase is a CAT tool that offers term extraction functionality, while Sketch Engine is a suite of professional lexical analysis tools that provides this facility as an extension to one of its core functions. Sketch Engine distinguishes more clearly than Phrase between single- and multi-word items. Phrase extracts lists of candidate terms as wordlists, while the Sketch Engine extraction tool works on the basis of keywords (see below). And finally, as a consequence of this, as we will see shortly, the results from Sketch Engine are generally far superior to those obtained from Phrase.

The main difference between a wordlist and a keyword list is that the former is based on simple word frequency, while the latter only contains items that are in some way characteristics of the texts that are being analysed (Brezina 2018: 79–80). The way this is achieved is by producing two separate wordlists, one from the texts that are being analysed (the 'focus corpus') and the other from another, usually much larger collection of texts (the 'reference corpus'), and then comparing the frequency of each item on the two lists. This comparison is used to derive the keyness score (Kilgarriff 2009).

Once a user's texts have been uploaded, they are compiled by the system. This process essentially entails adding part of speech tags using a sketch grammar and a term grammar to facilitate more powerful lexical analysis. Additionally, text in languages such as Chinese that do not include spaces between words is "segmented" in order to make the individual words and phrases accessible for lexical analysis.

Following the compilation, term extraction can be performed by selecting the Keywords function and configuring settings similar to those offered by Phrase. This will produce lists of single-word and multi-word items; according to the nomenclature used by Sketch Engine, the former are known as "keywords" and the latter as "terms". Using the same text as in Sect. 2 and selecting enTenTen13 as the reference corpus, the 1000-item list of keywords that is generated starts as set out in Fig. 3.

The first entries on the list of multi-word "terms" are as shown in Fig. 4.

From these lists, the general subject area is immediately clear, and anyone who possesses even a passing acquaintance with the subject matter of the text will recognise the relevance of many of the items that they contain. Of course, a complete list is generally much longer, and the relevance of items may quickly diminish as one searches further down it.

All in all, however, the quality of Sketch Engine's extraction is impressive, to say the least, and the fact that Sketch Engine bases this process on its keywords function—derived as it is from the contrasting word frequencies found in two separate corpora—means that they are much more accurate than those that Phrase derives from a simple wordlist as items that characterise the content of the text are foregrounded while most common lexical items are ignored.

When downloaded in XLS format, the resulting term list is as shown in Fig. 5.

Word	Word	Word
1 Keirsey •••	11 INTJ •••	21 ENTP •••
2 MBTI •••	12 ENFJ •••	22 Intuition •••
3 INFJ •••	13 ESTP •••	23 ENTJ •••
4 ISFP •••	14 INTP •••	24 Personalities •••
5 ESFJ •••	15 Sensing •••	25 introversion •••
6 Perceiving •••	16 Introversion •••	26 introvert •••
7 ISFJ •••	17 INFP •••	27 CPP •••
8 ESTJ •••	18 ESFP •••	28 INFJs •••
9 ISTP •••	19 Extraversion •••	29 Characteristics •••
10 ISTJ •••	20 ENFP •••	30 extraversion •••

Fig. 3 First thirty single-word "keywords" generated from the file mbti.html by Sketch Engine, with English Web 2013 (enTenTen13) as reference corpus

This can be sorted by one or other column, although by default the sorting is according to keyness score. However, in spite of this score, some items are clearly terms while others are items of the standard vocabulary that is needed in order to talk about this particular subject area. As before, the onus is still on the user to decide how many items from each of these categories to keep, and then to identify the correct equivalents in all probability using resources outside the tool.

For terminology work that is based on an entire field rather than a single text, it is possible for users to construct their own corpus semi-automatically from the web, either using their own texts or getting Sketch Engine to crawl the web for texts based on certain parameters that can be specified.

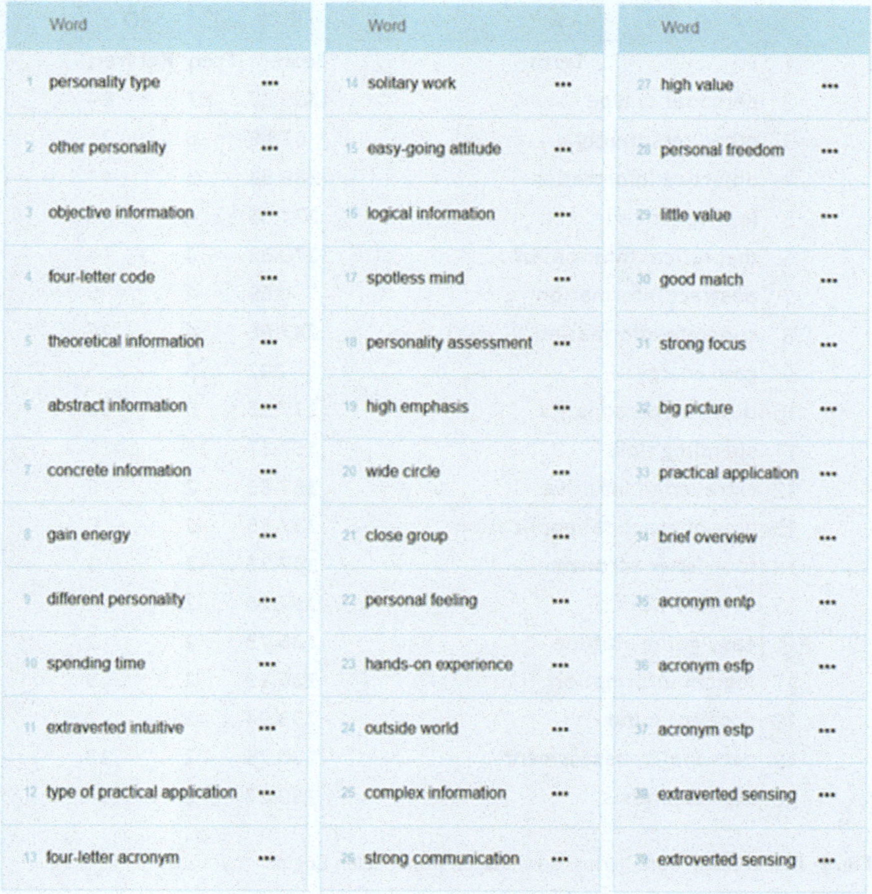

Word	Word	Word
1 personality type	14 solitary work	27 high value
2 other personality	15 easy-going attitude	28 personal freedom
3 objective information	16 logical information	29 little value
4 four-letter code	17 spotless mind	30 good match
5 theoretical information	18 personality assessment	31 strong focus
6 abstract information	19 high emphasis	32 big picture
7 concrete information	20 wide circle	33 practical application
8 gain energy	21 close group	34 brief overview
9 different personality	22 personal feeling	35 acronym enfp
10 spending time	23 hands-on experience	36 acronym esfp
11 extraverted intuitive	24 outside world	37 acronym estp
12 type of practical application	25 complex information	38 extraverted sensing
13 four-letter acronym	26 strong communication	39 extroverted sensing

Fig. 4 First 39 multi-word "terms" generated by Sketch Engine from the file mbti.html, with a sample of English Web 2013 as reference corpus

3.2 Bilingual Term Extraction

Unlike what we saw with monolingual terminology extraction, which is relatively easy to perform with any corpus software, bilingual terminology extraction is an altogether more specialised process. Bilingual corpora are far harder to locate or to construct than monolingual ones, and in addition, there are relatively few tools available that offer this functionality. Indeed, it seems that Sketch Engine may be one of the only major suites of corpus tools in which this process is explicitly possible. That said, compared to its monolingual extraction function, the results from its bilingual extraction are somewhat disappointing (although the extractions used to illustrate this section are admittedly based on small and medium-sized corpora).

	A	B	C	D
1	Term	Score	Freq	Ref freq
2	personality type	4399.83	87	84
3	other personality	378.2	6	21
4	objective information	289.51	5	44
5	four-letter code	274.65	4	0
6	theoretical information	270.33	4	4
7	abstract information	269	4	5
8	concrete information	247.66	4	26
9	gain energy	202	3	5
10	different personality	177.95	3	38
11	spending time	157.17	16	1419
12	extraverted intuitive	137.83	2	0
13	type of practical application	137.83	2	0
14	four-letter acronym	137.28	2	1
15	solitary work	137.28	2	1
16	easy-going attitude	136.73	2	2
17	logical information	136.19	2	3
18	spotless mind	133.94	2	7
19	personality assessment	130.76	2	13
20	high emphasis	130.27	2	14

Fig. 5 List of multi-word 'terms' downloaded from Sketch Engine

Sketch Engine employs corpora that have been tagged for part of speech, and the ABTE ("Automatic Bilingual Terminology Extraction") alignment algorithm that it uses works by calculating co-occurrence statistics for candidate terms on the basis of matching grammar rules (usually focusing on noun phrases) applied to parallel sets of monolingual terms that have been extracted separately (Baisa et al. 2015: 62–63). The Sketch Engine bilingual term extraction feature can be accessed by clicking on the Bilingual terms button on the Dashboard or else via its OneClick terms interface,[3] the difference being that with the latter route no permanent corpus is created. It needs parallel aligned text in order to function, and this can usually be most conveniently provided in the form of a TMX file, which can be easily exported from practically any CAT tool.

To continue along the same thematic lines as in previous sections, a TMX based on the topic of personality typology was imported into Sketch Engine. The file was of modest size (consisting of 8,306 English and 8,022 Traditional Chinese words), a

[3] https://terms.sketchengine.eu/.

fact that may have affected the quality of the bilingual pairings. By default, Sketch Engine used ententen20_tt31 and zhtenten17_simplified_stf2 as reference corpora.

Once a corpus has been created by uploading a TMX file and the Bilingual terms button has been clicked on, the user is taken to an interface that offers various listings, including single- and multi-word items in each language and what it calls "bilingual terms" (i.e. candidate term alignments), as presented in Fig. 6.

By default, items are arranged by L1 frequency, although they can be re-sorted according to a number of alternative measures. Sketch Engine suggests up to five candidate L2 equivalents for each L1 term, and the user can select the one he or she considers most appropriate. In the event, most of the pre-selected equivalents are incorrect (being generally partial matches or linked in some way to the correct

Fig. 6 The first eight of 100 "bilingual terms" produced by Sketch Engine on the subject of personality typology

Source term	Target term
personality type	人格 性格 人格 型態 型態 Personality
endif	if J. Butt 圖書館 助理 X檔案
type indicator	指標 Indicator Myers- Briggs Briggs 類型
introversion	與直覺 與內 外向 與內 與 情感 與
extraversion	外向 與內 與 情感 與直覺 與內 與
four-letter	代表 內向 縮寫 字母 縮寫 字母 代碼 代碼 代表
temperament sorter	性格 分類 分類 方法 分類 創作者 心理 學家
personality	性格 Personality 人格 人格 型態 型態

```
-<termEntry id="99">
  -<langSet xml:lang="en">
    -<tig>
       <term>personality type</term>
       <termNote type="count">14</termNote>
       <termNote type="status">new</termNote>
       <termNote type="example"/>
       <termNote type="toknum">-1</termNote>
       <termNote type="corpus">user/markshut/ukmpdby_en</termNote>
       <termNote type="cofreq">19</termNote>
       <termNote type="score">12.398</termNote>
     </tig>
   </langSet>
  -<langSet xml:lang="zh-Hans">
    -<tig>
       <term>人格 型態</term>
       <termNote type="count">66</termNote>
       <termNote type="example"/>
       <termNote type="toknum">-1</termNote>
       <termNote type="corpus">user/markshut/ukmpdby_zh_hans</termNote>
     </tig>
   </langSet>
 </termEntry>
```


Fig. 7 termEntry element of the TBX file downloaded from Sketch Engine

one), although, in some cases, a better alternative than the one suggested is available to select. When a user chooses a different equivalent, this will be included in the downloaded TBX file; thus if the equivalent of the first item, "personality type", is corrected from 人格 ("personality") to 人格型態 ("personality type"), it gives rise to the TBX (Term Base eXchange) termEntry element shown in Fig. 7.

Trying a similar experiment with the larger infopankki parallel corpus, consisting of 189,179 English and 216,727 French words and downloaded from the Opus website (see below), we get the tentative initial alignments displayed in Fig. 8.

As can be seen, the corpus focuses on information on the subject of emigrating to Finland. Once again, in spite of the larger size of the corpus, the quality of most of these candidate matches is not high, with no correct equivalent even being offered for the last item, 'finnish', for example.

Space does not permit a detailed consideration of this, but cases where no correct pairing of terms is offered can be further investigated via Sketch Engine's Parallel Concordance search function. However, since the process of creating bilingual corpora in Sketch Engine is less flexible than that for monolingual ones, we need to investigate alternative methods for assembling large amounts of parallel data. This will form the topic of the next section.

Fig. 8 The first eight of 100 candidate term alignments produced by Sketch Engine on the basis of a larger corpus on the subject of emigrating to Finland

Source term	Target term
residence permit	**séjour** × permis demande finlande prolongation
finnish immigration	**migri** × immigration office national guichet immigration migri
upper secondary school	**formation préparatoire** × étude lycéenne lycéenne formation professionnelle luva
health centre	**centre médical** × centre médico-social permanence propre centre infirmier
te office	**activité économique** × demandeur d demandeur chômage recherche d
finnish social security	**sociale finlandaise** × sécurité sociale sécurité sociale finlandaise finlandaise page sécurité sociale
service point	**guichet** × migri office national immigration pt service
finnish	**russe** × arabe somali kt kurde

4 Acquiring Parallel Text

This section is intended to offer a possible solution to one of the problems that were raised at the beginning and then again at the end of Sect. 3.2, namely that it is much harder to locate or construct a bilingual corpus than a monolingual one. Needless to say, the amount of available bilingual aligned data is dwarfed by the volume of monolingual data that can be easily accessed, besides which the latter is far easier to gather and process—although generally speaking not nearly as useful in the context of translation technology. Indeed, it is difficult to overemphasise the importance of parallel aligned text as a resource that drives the advance of translation technology, so by helping our students to access it online, we are doing them a significant favour, not least because as learners of translation technology students often hear about the advantages of having large TMs but then find themselves restricted to very small, starter ones. The purpose of this final section is thus to present some resources from which it is possible to download bilingual aligned data in many different language pairs as well as to consider the uses to which this can be put in the context of translation technology teaching.

There are a number of reasons why it would make sense to include this in a translation technology curriculum, although to my knowledge, at the time of writing it does not feature in a significant number of courses. In spite of the problem stated above, there is in fact a considerable amount of parallel data that is available for download on the internet, if one only knows where to find it. Section 4.1 focuses on the OPUS website, which is a particularly rich source of bilingual data, while a number of alternative resources form the topic of Sect. 4.2. OPUS and some of the other resources are free of charge and only contain data that is in the public domain; however, other sites charge for their content. What the course designers choose to do with this data is of course up to them, although three obvious uses might be quickly growing a TM in order to enable students to experience something approaching the full potential of CAT tool technology, enabling systematic term extraction through the analysis of large parallel corpora, and adding training data to an MT engine that is under construction.

4.1 OPUS—An Open Source Parallel Corpus

The OPUS website, which is curated by Professor Jörg Tiedemann of the University of Helsinki, describes itself as "a growing collection of translated texts from the web" (OPUS 2021a). All data is in the public domain and freely downloadable; as new datasets are added, they are converted, aligned, linguistically tagged and saved in a wide range of formats (OPUS 2021a). Various domains are covered by the data, the most prominent being the political (chiefly represented by data from the European Parliament), movie subtitles, localised open-source software, news, religious texts and data from Wikimedia projects (Tiedemann 2016: 384). New data is constantly being added; for example, four sets were uploaded from January to October 2021 (the time of writing). As of 2020, OPUS comprised corpora covering over 700 languages and language variants that together constituted more than 70,000 aligned text pairs (Aulamo et al. 2020: 3782). At the time of writing, the site offers a total of 69 datasets ranging from as few as around 860 sentence fragments right up to 7.37 bn. The precise number of languages cannot be directly accessed, although some individual resources include as many as several hundred (e.g. 244 for Ubuntu, 374 for Tatoeba and more than 500 for MT560). While most corpora are multilingual, around 17 relate to a single language pair. Data from OPUS has been used by a number of other projects, including, for example, CASMACAT, Reverso and Sketch Engine (OPUS 2021a).

corpus	doc's	sent's	en tokens	zh tokens	XCES/XML	raw	TMX	Moses	mono	raw
CCMatrix v1	1	71.4M	1.1G	108.0M	xces en zh	en zh	tmx	moses	en zh	en zh
WikiMatrix v1	1	2.6M	1.0G	67.0M	xces en zh	en zh	tmx	moses	en zh	en zh
UNPC v1.0	94770	14.2M	379.8M	36.3M	xces en zh	en zh	tmx	moses	en zh	en zh
wikimedia v20210402	1	0.2M	349.2M	7.9M	xces en zh	en zh	tmx	moses	en zh	en zh
MultiUN v1	67167	8.0M	220.8M	13.5M	xces en zh	en zh	tmx	moses	en zh	en zh
XLEnt v1.1	1	6.0M	19.5M	17.3M	xces en zh	en zh	tmx	moses	en zh	en zh
Tanzil v1	30	0.2M	5.6M	1.7M	xces en zh	en zh	tmx	moses	en zh	en zh
bible-uedin v1	4	0.1M	3.1M	3.4M	xces en zh	en zh	tmx	moses	en zh	en zh
UN v20090831	1	74.1k	3.7M	1.2M	xces en zh	en zh	tmx	moses	en zh	en zh
TED2013 v1.1	1	0.2M	3.1M	0.9M	xces en zh	en zh	tmx	moses	en zh	en zh
News-Commentary v16	7107	69.0k	3.2M	89.7k	xces en zh	en zh	tmx	moses	en zh	en zh
QED v2.0a	176	20.8k	0.3M	0.3M	xces en zh	en zh	tmx	moses	en zh	en zh
PHP v1	3274	45.0k	0.5M	0	xces en zh	en zh	tmx	moses	en zh	en zh
TED2020 v1	173	16.2k	0.3M	70.6k	xces en zh	en zh	tmx	moses	en zh	en zh
WMT-News v2019	8	12.0k	0.3M	20.2k	xces en zh	en zh	tmx	moses	en zh	en zh
SPC v1	1	2.2k	69.2k	61.3k	xces en zh	en zh	tmx	moses	en zh	en zh
EUbookshop v2	18	0	0.1M	0	xces en zh	en zh	tmx	moses	en zh	en zh
infopankki v1	256	8.9k	93.8k	17.8k	xces en zh	en zh	tmx	moses	en zh	en zh
tico-19 v2020-10-28	1	3.1k	70.4k	10.4k	xces en zh	en zh	tmx	moses	en zh	en zh
ELRC_2922 v1	1	0.1k	2.9k	0.8k	xces en zh	en zh	tmx	moses	en zh	en zh
Ubuntu v14.10					xces en zh	en zh	tmx	moses	en zh	en zh
total	**172992**	**103.2M**	**3.1G**	**257.7M**	**103.2M**		**101.3M**	**101.4M**		

Fig. 9 List of aligned text resources available for the English <> Chinese language pair from the OPUS website

The OPUS interface[4] thus permits users to download considerable amounts of parallel data in a wide range of language pairs. By specifying source and target languages, and also resource size if desired, it is possible to see a list of available corpora, as shown in Fig. 9.

As demonstrated in Fig. 9. 21, resources are available for download for English <> Chinese. However, this figure covers all varieties of both languages, while a user would very likely need to narrow the search to either Traditional or Simplified Chinese and/or British or American English, which reduces the range of available possibilities. The table in Fig. 9 also indicates the size of each resource in terms of number of documents, sentences and tokens in each of the two languages and the availability of downloads in a large number of formats. Users can click on the link of a particular resource to be taken to it. The coloured highlighting is designed to provide a visualisation of the size of the resource according to the combined number of source and target tokens (OPUS 2021a), green (the items near the top) indicating more than two million (Tiedemann, personal e-mail, 25 October 2021).

The number of resources varies greatly between language pairs as indicated in Table 1.

These pairs were selected more or less at random and indicate in general that, while there are generally more resources available for pairs involving major Western languages (whether or not English forms part of the pair), other languages that have perhaps not usually been associated with the availability of large amounts of digital data are also quite well represented. (The paucity of data available for English <> Japanese seems to be anomalous.) As stated above, most of the resources listed in each case are multilingual. The range and quantity of data that is available is of

[4] This is available at http://opus.nlpl.eu/.

Table 1 Availability of bilingual data on the OPUS site for a set of sample language pairs (ignoring language variants in each case)

French <> English 54
German <> English 45
French <> German 40
Finnish <> English 35
English <> Korean 19
Russian <> Chinese 17
English <> Swahili 16
French <> Swahili 15
Persian <> Ukrainian 12
Korean <> Swahili 11
English <> Japanese 1

course good news for those who teach translation technology to multilingual groups (a very common practice in some parts of the world) as no participants will be likely to be excluded from an activity because of a lack of data in their language pair. In this respect, a multilingual resource, such as CCMatrix, is excellent for providing data for language combinations that are not so frequently opted for. Finally, a small number of resources offer intralingually aligned alternative translations, as discussed briefly below.

Most, if not all, resources can be downloaded as TMX files, compressed using the .GZ format. Although many students may be unfamiliar with this file format, it can be handled very simply using a free file extraction utility (7-zip works well, for example). Many of the resources, including some of those listed in Fig. 9, are extremely large: CCMatrix has 71.4 m sentences, for example. This would probably be too big for teaching purposes, and would take a long time to import, but at a pinch, students could be shown how to extract a more manageable subset by editing the TMX file directly (or the teacher could do this on their behalf) using a tool such as Goldpan.[5] Other resources towards the bottom end of those listed in Fig. 9 are much more modestly sized, and a TMX consisting of a few tens or even hundreds of thousands of sentences would be more than enough to convert a small starter TM of a few dozen or a couple of hundred segments into a large-sized bilingual resource capable of fulfilling the promise of Pierre Isabelle's dictum that was quoted above.

Once the downloaded data has been extracted, it can be easily imported into a TM using the CAT tool's standard procedures. Once this has been done, students will, for example, be able to receive useful and meaningful results from a concordance search, a procedure that is in many ways very similar to carrying out a Linguee.com enquiry but from within a CAT tool and focused on material that may be highly relevant to the text that is being translated. In the two figures below, the OPUS Wikimedia English-Russian corpus consisting of 312,152 segments has been imported into Phrase and is being used for different concordance searches. The search illustrated in Fig. 10 demonstrates how *oral* appears to be a possible translation equivalent for the Russian medical term *peroralny*.

[5] https://logrusglobal.com/goldpan.html.

Fig. 10 A concordance search conducted in Phrase to find an English equivalent for the Russian term *peroralny* contained in a sentence for translation

1	В другом исследовании с участием 20 мг однократной пероральной дозы у здоровых людей, период полураспада среднем 5,6 часа.[1][2]	9	In another study involving a 20 mg single oral dose in healthy men, the elimination half-life averaged 5.6 hours.[1][2]
2	Он был одобрен комитетом продовольствия и медикаментов (FDA) в ноябре 2005 года.[2][4] По данным FDA (Май 2007 года), почечная недостаточность и цитопений были зарегистрированы у больных, получающих деферазирокс таблетки пероральной суспензии.	5	It was approved by the United States Food and Drug Administration (FDA) in November 2005. [2][4] According to FDA (May 2007), renal failure and cytopenias have been reported in patients receiving deferasirox oral suspension tablets.

In Fig. 11, on the other hand, another search yields 50 hits offering various equivalents for the Russian noun *kompleks*, which is not terminological in nature but is a standard item of vocabulary that can sometimes be problematic to translate.

Such extensive search results clearly cannot be achieved using a starter TM, but by importing a sizeable TMX file, students can experience the real value of important CAT tool features that might otherwise not be fully appreciated.

Out of the 69 resources currently available, the following twelve are perhaps worth a special mention either because they exemplify a certain category of OPUS data or by virtue of the particular significance for teaching that they possess:

Fig. 11 A concordance search conducted in Phrase to find possible equivalents for the Russian word *kompleks*

10	Жилой комплекс «Интерлейс»	33	The Interlace
11	Храмы, входящие в комплекс:	32	The temples along the hiking track among others are:
12	Комплекс не армирован бетоном.	28	The total length of the complex is 1,090 m (3,025 m2, 7,562 m3). [132][133] It is not reinforced by concrete.
13	Комплекс занимает 52 тысячи м2[8]	28	The complex covers about 52,000 square metres (560,000 sq ft).[8]
14	Комплекс Фенна-Мэтьюса-Ольсона	27	Fenna-Matthews-Olson complex

- OPUS-100 is described as "an English-centric multilingual corpus" that covers one hundred languages and was randomly sampled from existing corpora within the project (OPUS 2021b). Users can download either the entire corpus or else individual language pairs, each of which includes English as either source or target language. The entire corpus is composed of approximately 55m sentence pairs, with the data available in individual language pairs ranging from more than one million sentence pairs to fewer than ten thousand (ibid.). The individual language pairs feature in Sketch Engine as a series of 40 "OPUS2" parallel corpora. OPUS-100 could be considered to be quite a good general-purpose parallel corpus, while the amount of data available for some language pairs need not be intimidating as some subsets are available for download.
- Tatoeba is a collection of translated sentences from the Tatoeba website,[6] a site offering crowdsourced translations of both simple and more complex sentences mostly relating to spoken language and intended as a resource for language learning. Available in 374 languages and consisting of a total of 10.24m sentence fragments, if not the largest available for download from the OPUS site, this corpus contains data in more languages than almost any other. Tatoeba data contains much conversational language and, consequently, it is also likely to include a large number of interrogative sentences; this might be useful when training an MT engine as this sentence type is often poorly represented in much training data (see Shen 2010).
- MT560 is a vast machine translation dataset that covers over 500 languages. Described as a "many-to-English machine translation dataset" (OPUS 2021c), it is chiefly intended to facilitate language technology research in low resource languages (OPUS 2021c). Like a number of other OPUS resources, it must be downloaded from an independent website.[7]
- Europarl is a set of parallel data extracted from the proceedings of the European Parliament. It includes data in 21 languages and contains a total of 30.32m sentence fragments. The amount of data available in individual language pairs ranges from 0.4m to 2.1m sentences. The most recent release is from 2012.
- OpenSubtitles is the latest in a series of collections of translated film subtitles from Opensubtitles.org,[8] previous datasets having been created in 2011, 2012, 2013 and 2016. It contains data in 62 languages and totals 3.35bn sentence fragments. What is interesting about this resource is that it now also includes intralingual alignments of alternative translations (Tiedemann 2016: 384). On the other hand, given the nature of the subtitling process, it is possible that a proportion of target segments are more condensed compared to their source equivalents.
- UNPC (the United Nations Parallel Corpus) contains 172.04m sentence fragments of manually translated material from the years 1990–2014 (Ziemski et al. 2016: 1) in the six official languages of the UN (Arabic, Chinese, English, French, Russian and Spanish). Data for each language pair ranges from 14.2 to 30.3m sentences.

[6] http://tatoeba.org/.

[7] http://rtg.isi.edu/many-eng/.

[8] http://www.opensubtitles.org/.

- Wikimedia contains 31.62m sentence fragments of Wikipedia translations in 306 languages, which is presumably the total number of languages that had an edition of the encyclopaedia when the corpus was created. A much larger amount of similar data is available in WikiMatrix, although for a smaller number of language pairs.
- The Common Crawl[9] is represented on the website by CC Aligned, which contains parallel data totalling 2.25 bn sentence fragments in 113 languages. The initial alignments were with English, although the recognition that many English documents were aligned with documents in multiple other languages has led to the creation of some non-English aligned document pairs.
- ParaCrawl and MultiParaCrawl were both extracted from the ParaCrawl project.[10] ParaCrawl contains 3.58bn sentence fragments relating to 39 languages, with nearly all language pairs including English. MultiParaCrawl, on the other hand, offers 1.19 bn sentence fragments across 38 languages, mostly in language pairs that do not include English. The two datasets only include EU languages.
- Books is a collection of books that are no longer in copyright and is intended for personal, educational and research purposes. The dataset covers 16 languages and totals 0.91 m sentence fragments, making it one of the smaller resources available through OPUS. The titles included in the data are listed on the website of Farkas Translations.[11]
- The ELRC (European Language Resource Coordination) public datasets are a collection of 150 corpora covering topics that range from annual reports, statistical datasets and lists of goods to banking, legislation and insurance, and also including a number of thematically unspecified parallel corpora. No overall total numbers of languages or sentence fragments are provided as each dataset is linked to separately. Each individual resource typically relates to two or more languages and is a few thousand or tens of thousands sentence fragments in size.
- Ubuntu is one of OPUS's datasets of localised open-source software. It is a modestly-sized parallel corpus consisting of localisation files from the Ubuntu operating system. Although the corpus includes data in 244 languages, it contains only 7.73m sentence fragments in total, with typically up to a few thousand words in each language pair.

Finally, the site also provides links to a number of tools that are connected to the project. Among these is OPUS-CAT,[12] which is a Windows-based neural MT system that can be run in a local, secure manner via CAT tool plugins. This has clear pedagogical potential that could be usefully explored in a number of contexts.

Taken together, the corpora and other resources discussed above quite clearly represent an excellent set of resources that can greatly enrich students' learning experience and facilitate a much more professionally realistic use of different types

---

[9] https://commoncrawl.org/.

[10] http://paracrawl.eu/download.html.

[11] https://farkastranslations.com/bilingual_books.php.

[12] https://helsinki-nlp.github.io/OPUS-CAT/.

of translation technology. Of all the download formats, TMX is one of the simpler ones and has the advantage that it can be imported into an existing TM immediately after extraction. Some of the resources listed above are extremely large, and although some of them might be derived from multiple documents, it is not generally possible to download these as individual files as they are all combined within a single .GZ archive. On the other hand, some resources are much smaller (e.g. Wikimedia, Books and Ubuntu), or allow the user to download much more modestly-sized individual datasets (e.g. ELRC). This means that some datasets potentially offer a suitable way of bringing about a modest enlargement of a starter TM, if that is what the teacher wants to achieve, perhaps as a first step.

4.2 Lists of Other Parallel Data Resources

As stated above, the reason why OPUS has been focused on to such an extent is that a considerable amount, and wide variety, of freely downloadable data has been gathered together in one place. It could then be argued that there is little point in going in detail into resources other than the OPUS website, particularly given the considerable overlap that sometimes exists with resources that are available there. However, it is probably worth at least briefly mentioning a number of others, not least because of the additional data that can sometimes be located there.

The following are some alternative sites that offer links to free parallel data.

- As of October 2021, the Wikipedia article on Parallel Text[13] offers links to some sixteen external resources. One of these is the OPUS website, two more (JRC and Europarl) are linked to from there, while around 7–8 others also contain parallel text datasets. The remaining ones provide interfaces for non-downloadable resources or other similar services.
- The Moses Links To Corpora page[14] provides ten links to multilingual and three to bilingual resources. Of the former, the OPUS site and several resources held on it are linked to, as well as a News Commentary corpus, Microtopia, which is a corpus of data in 11 languages extracted from Twitter and Sina Weibo, and an Asian Scientific Paper Excerpt Corpus.
- The OSCAR page[15] lists resources in 170 languages ranging from one kilobyte to nearly three terabytes in size, which together form part of the OSCAR corpus (or Open Super-large Crawled Aggregated coRpus). While the data included on this site appears to be unique, it is, however, unclear whether the corpus contains any aligned data. Access to the downloadable datasets must be requested via e-mail.

[13] https://en.wikipedia.org/wiki/Parallel_text.

[14] http://www.statmt.org/moses/?n=Moses.LinksToCorpora.

[15] https://oscar-corpus.com/.

- Finally, the CLARIN ERIC (Common Language Resources and Technology Infrastructure European Research Infrastructure Consortium) page[16] provides links to 47 bilingual and 39 multilingual corpora. Of the latter, five contain text in more than 50 languages. Most of the resources appear to be unique data resources in a considerable number of languages, both European and non-European.

Besides these, there are a number of free tools, such as Bitextor,[17] that are intended for harvesting TMs from bilingual websites, but these lie beyond the scope of the present paper. Although somewhat different in nature, it is also worth mentioning that MyMemory[18] allows users to download a TM that is created on the fly according to the specifications that they enter. Downloads, however, can vary greatly in size, from a thousand or more segments down to little more than a dozen.

The fact remains that there is some degree of duplication of sources from one site to another, and although there is a large amount of 'general purpose' data, you will be lucky indeed if you find data that fits closely with a particular subject area or job specification. The picture when we look at sites that offer data for payment is rather different as these sites tend to host significant amounts of often more targeted aligned data that is not available from other sources.

- Probably the best-known site for purchasing datasets is the Netherlands-based TAUS (Translation Automation User Society).[19] TAUS does currently offer a Corona Crisis Corpus containing data in six language pairs relating to virology, epidemics, medicine and healthcare for free download. However, apart from this, all data obtained from the site needs to be purchased, and TAUS has three models for this. Firstly, the data library provides off-the-shelf MT training data. Secondly, under the matching data service, a customer submits a typical sample of their data that they would like to see matched. TAUS then runs an algorithm to identify data from their repository that best matches the sample, after which the customer can pay and download it. Thirdly, the Data Marketplace allows customers to purchase specific datasets that are being marketed by other members. The platform also permits members to monetise their own data by uploading and marketing it. There are three levels of membership, depending on a user's expected level of ongoing data needs, while data is marketed at prices ranging from €0.0035 to €0.0065 per word. In spite of the abundance of data—there are in excess of 1500 data types listed in the marketplace—there is no guarantee that a user will be able to find exactly what he or she is looking for. For universities, TAUS offers an academic programme that provides access to the data matching service.
- A number of other sites also provide data for sale. For example, the Linguistic Data Consortium hosted at the University of Pennsylvania[20] offers members and non-members access to a catalogue of hundreds of corpora in a range of language

[16] https://www.clarin.eu/resource-families/parallel-corpora.

[17] https://sourceforge.net/projects/bitextor/.

[18] https://mymemory.translated.net/.

[19] https://www.taus.net/.

[20] https://catalog.ldc.upenn.edu/.

pairs and subject areas. A typical item may cost up to several thousand dollars for a non-member to download.

In terms of where each of the course components described in Sects. 2–4 should be placed, it seems that curriculum designers can allow themselves a reasonable amount of flexibility. Regarding term extraction using Phrase or another CAT tool, it seems that it might be most appropriate to include this at the same time that the other features of the tool are being taught. Sketch Engine can be picked up by most students fairly rapidly, so that component can either be included at the same time or perhaps later on, as term extraction may possibly be considered a less central topic than a number of others. In a sense, the sooner students can start to benefit from large translation memories the better, although once again it may be considered more appropriate for certain other topics to be included first.

5 Conclusion

The article sets out to discuss two topics within translation technology teaching, one of which is well established and the other probably not yet very well represented. In Sects. 2 and 3, a brief overview of the possibilities for extracting terminology that are offered by various CAT tools, the Phrase term extraction function was examined in detail. While this has certain limitations, it potentially serves the useful purpose of introducing students to the concept of term extraction from within an tool with which they are already familiar. Not all CAT tools offer this feature, however, and when it comes to bilingual term extraction, the options become highly limited. Once the Phrase feature has been tried out, however, it is recommended that students be introduced to a system such as Sketch Engine that produces much better results, for monolingual if not for bilingual extraction, and will allow them to become more closely acquainted with the processes that are involved.

The discussion of possible sources of data for allowing students to work with larger TMs that was presented in Sect. 4 was more exploratory in nature. It has shown that substantial online sources of parallel text exist, has made some initial recommendations as to which might be particularly useful, and has also offered suggestions as to how the datasets that are on offer might be exploited in the context of a translation technology course, a topic that it is believed is relatively untried.

This article has described the tools as they appeared in the early 2020s. However, starting from the end of 2022, a wave of new commercial AI applications has broken over us, leaving many wondering what the way forward should be. Will all the tools we cover with our students cease to exist? Should we throw away our teaching materials and start again? Is there even a need for our specialism any more? While the new technology represents a major challenge for us—with its appearance likely to prove at least as significant as the emergence of the world-wide web in the early 1990s—I believe the answer to all these questions is no. That said, for the here and now, we are faced with the urgent task of equipping our students with the skills they need in order to thrive in a much-changed professional landscape. The emergence of MT over the last decade greatly increased the need for translation—and also radically altered the nature of what many translation professionals do—and there is no reason to believe that the phenomenal rise of AI-driven technology will be any different, as it will change the nature of the translation industry in ways we cannot perhaps yet imagine. It is our responsibility to equip our students to become skilled users of this new technology.

While change is likely to be rapid, the tools we have been teaching our students are not going to disappear overnight, although we can currently only guess at what they will look like in as little as 1–2 years' time. Essentially, at least for the short and medium terms, they are likely to remain substantially the same, but perhaps with the incremental addition of AI-driven capabilities and so able to achieve significantly more—like the current tools on steroids, perhaps. For now, however, we may at least venture predictions about the shape that these new capabilities will take. While I will leave it to others who are more knowledgeable than me to discuss the potential for corpus linguistics in general, for the purposes of the present article, instead of setting parameters as presently happens, it is not difficult to visualise providing an AI-enabled system with a (possibly extremely detailed) prompt to get it to produce a nuanced bilingual term extraction or perform a targeted parallel data acquisition operation. Performance in both these tasks is likely to be excellent, although the place they will occupy in an AI-enabled workflow is less clear maybe. In the case of CAT tools, too, the technology will not disappear but its reach will be greatly extended.

As for our teaching materials, it is clear that they will need to undergo a significant evolution in the short and medium terms. This is likely to start with supplementing our existing courseware with guidance on how to use non-specialised AI technology for translation purposes while we wait for existing translation tools to catch up, and at the same time creating teaching materials for new tools—and significantly updated versions of existing tools—as they are developed.

The need at present is to proceed with confidence, even if the next steps may seem somewhat unclear.

References

Aulamo, Mikko, Umut Sulubacak, Sami Virpioja, and Jörg Tiedemann. 2020. OpusTools and Parallel Corpus Diagnostics. In *Proceedings of the 12th Conference on Language Resources and Evaluation (LREC 2020)*. Marseille, 11–16 May 2020. 3782–3789. https://aclanthology.org/2020.lrec-1.467.pdf. Accessed 27 May 2023.

Baisa, Vít, Barbora Ulipová, and Michal Cukr. 2015. Bilingual Terminology Extraction in Sketch Engine. *Ninth Workshop on Recent Advances in Slavonic Natural Language Processing*. The Czech Republic, December 2015. 61–67. http://www.sketchengine.eu/wp-content/uploads/Bilingual_Terminology_Extraction_2015.pdf. Accessed 27 May 2023.

Bondi, Marina, and Mike Scott. Eds. 2010. *Keyness in Texts*. Amsterdam & Philadelphia: Benjamins.

Brezina, Vaclav. 2018. *Statistics in Corpus Linguistics: A Practical Guide*. Cambridge: Cambridge University Press.

Champagne, Guy. 2004. The Economic Value of Terminology: An Exploratory Study. Submitted to the Translation Bureau of Canada April 30, 2004. https://www.danterm.dk/docs/EconomicValueTerminology-1.pdf. Accessed 27 May 2023.

Chan, Venus, and Mark Shuttleworth. 2023. Teaching Translation Technology. In *The Routledge Encyclopedia of Translation Technology*, 2nd ed, ed. Sin-wai Chan, 259–279. Abingdon and New York: Routledge.

Ghislandi, Massimo. 2014. A Major Milestone—100,000 Studio Licences Sold! Posted April 23 2014 in *Trados Blog*. https://www.trados.com/blog/a-major-milestone-100000-studio-licences-sold.html. Accessed 27 May 2023.

Isabelle, Pierre, Marc Dymetman, George Foster, Jean-Marc Jutras, Elliott Macklovitch, François Perrault, Xiaobo Ren, and Michel Simard. 1993. Translation Analysis and Translation Automation. In *Proceedings of the Fifth Conference on Theoretical and Methodological Issues in Machine Translation of Natural Languages*. July 14–16, Kyoto, Japan, 201–17. https://aclanthology.org/1993.tmi-1.17/. Accessed 27 May 2023.

Kilgarriff, Adam. 2009. Simple maths for keywords. In *Proceedings of Corpus Linguistics Conference CL2009*. University of Liverpool, UK, July 2009, eds. M. Mahlberg, V. González-Díaz, and C. Smith. https://www.sketchengine.eu/wp-content/uploads/2015/04/2009-Simple-maths-for-keywords.pdf. Accessed 27 May 2023.

Korkontzelos, Ioannis, and Sophia Ananiadou. 2022. Term Extraction. In *The Oxford Handbook of Computational Linguistics*, 2nd ed, ed. Ruslan Mitkov. New York: Oxford University Press, 991–1012. https://doi.org/10.1093/oxfordhb/9780199573691.013.004

Nimdzi Insights. 2021. Memsource. Updated 4 October 2021. https://web.archive.org/web/20211016130133/https://www.nimdzi.com/tms/memsource/. 21 May 2023.

OPUS. 2021a. OPUS—An Open Source Parallel Corpus. https://opus.nlpl.eu/. Accessed 27 May 2023.

OPUS. 2021b. OPUS-100 Corpus. https://opus.nlpl.eu/opus-100.php. Accessed 27 May 2023.

OPUS. 2021c. MT560—A Many-to-English Machine Translation Dataset. https://opus.nlpl.eu/MT560.php. Accessed 27 May 2023.

Phrase. 2023. Phrase (Frm. Memsource)—Localization & Translation Software. https://phrase.com/. 21 May 2023.

Rodríguez-Castro, Mónica. 2018. An integrated curricular design for computer-assisted translation tools: Developing technical expertise. *The Interpreter and Translator Trainer* 12 (4): 355–374. https://doi.org/10.1080/1750399X.2018.1502007.

Rothwell, Andrew, and Tomáš Svoboda. 2019. Tracking translator training in tools and technologies: Findings of the EMT survey 2017. *The Journal of Specialised Translation* 32: 26–60. https://jostrans.org/issue32/art_rothwell.pdf. Accessed 27 May 2023.

RWS Documentation Center. 2021. Creating AutoSuggest dictionaries. https://docs.rws.com/813470/566102/sdl-trados-studio-2021/creating-autosuggest-dictionaries. Accessed 27 May 2023.

Shen, Ethan. 2010. Comparison of Online Machine Translation Tools. https://web.archive.org/web/20140324051506/http://www.tcworld.info/e-magazine/translation-and-localization/article/comparison-of-online-machine-translation-tools/. Accessed 27 May 2023.

Tiedemann, Jörg. 2016. OPUS—Parallel Corpora for Everyone. *Baltic Journal of Modern Computing* 4:2, Special Issue: Proceedings of the 19th Annual Conference of the European Association of Machine Translation (EAMT). 384. https://www.bjmc.lu.lv/fileadmin/user_upload/lu_portal/projekti/bjmc/Contents/4_2_28_Products.pdf. 27 May 2023.

Wright, Sue Ellen, and Gerhard Budin. 1996. *Handbook of Terminology Management, Volume 1: Basic Aspects of Terminology Management*. Amsterdam & Philadelphia: Benjamins.

Zhang Xiaochun, and Lucas Nunes Vieira. 2021. CAT teaching practices: an international survey. *The Journal of Specialised Translation* 36a: 99–124. https://www.jostrans.org/issue36/art_zhang.pdf. Accessed 27 May 2023.

Ziemski, Michał, Marcin Junczys-Dowmunt, and Bruno Pouliquen. 2016. The United Nations Parallel Corpus v1.0. *Language Resources and Evaluation (LREC'16)*. Portorož, Slovenia, May 2016, 1–5. https://conferences.unite.un.org/UNCorpus/Content/Doc/un.pdf. Accessed 27 May 2023.

Mark Shuttleworth has been involved in translation studies research and teaching since 1993, at the University of Leeds, Imperial College London, University College London and, most recently, Hong Kong Baptist University. His publications include the *Dictionary of Translation Studies*, as well as articles on translation technology teaching, metaphor in translation, translation and the web, and Wikipedia translation. The use of digital methodologies in translation studies research is also an interest of his. His monograph *Studying Scientific Metaphor in Translation* was published in 2017 and he is currently finalising a second edition of the *Dictionary*.

How Do Students Perform and Perceive Parallel Corpus Use in Translation Tasks? Evidence from an Experimental Study

Kanglong Liu, Yanfang Su, and Dechao Li

1 Introduction

The compilation of corpora, utilisation of corpus tools, and application of corpus evidence for translational decisions are widely recognised as fundamental components of translation competence (Varantola 2003). Amongst various types of corpora, parallel corpora have emerged as the most valuable and effective resources, providing direct translation solutions for translators (Liu 2020). Extensive research has demonstrated the usefulness of parallel corpora for student translators, enabling them to extract desired terminology or concordances (Santos and Frankenberg-Garcia 2007), observe expert translators' approaches to translation problems (Monzó Nebot 2008), and explore potential information loss or supplementation during the translation process (Pearson 2003). Parallel corpora are believed to significantly enhance the competence and confidence of translation trainees (Zhu and Wang 2011). However, there is a lack of longitudinal and empirical research evaluating the effectiveness of corpus use in translator training (Frérot 2016). Previous studies have mainly focused on conceptual discussions, emphasising the advantages of corpus-assisted translation. Thus, further experimental research is necessary to evaluate the efficacy and limitations of employing parallel corpora in translation classrooms.

K. Liu · Y. Su · D. Li (✉)
The Hong Kong Polytechnic University, Hong Kong, China
e-mail: dechao.li@polyu.edu.hk

K. Liu
e-mail: kl.liu@polyu.edu.hk

Y. Su
e-mail: yanfang.su@connect.polyu.hk

J. Pan and S. Laviosa (eds.), *Corpora and Translation Education*, New Frontiers in Translation Studies, https://doi.org/10.1007/978-981-99-6589-2_7

2 Related Work

2.1 *Types of Corpora in Corpus-Assisted Translation Teaching*

Corpus-assisted translation teaching, an interdisciplinary approach situated at the intersection of corpus linguistics, translation studies, and educational theories (Bernardini 2004), encompasses the purposeful use of corpus tools and data to enhance translation instruction. This approach draws upon the methodology of corpus linguistics, employing a bottom-up approach wherein students systematically analyse and interpret corpus data to derive meaningful insights (Boulton and Cobb 2017). It also incorporates elements of descriptive translation studies, such as the examination of translation universals within the classroom setting (Laviosa 2008). The design of corpus-assisted translation training syllabi incorporates diverse educational approaches, including discovery learning, project learning (Bernardini 2016), and the task-based approach (Marco and Van Lawick 2009). The interdisciplinary nature of corpus-assisted translation teaching underlies its numerous merits, which have garnered significant scholarly attention over the past two decades (Biber et al. 1998). One of the key merits of corpus-assisted translation teaching lies in its inductive and student-centred approach to learning. By utilising corpora, translation students gain access to a vast collection of authentic texts (Bowker 2002) and are encouraged to actively engage as language researchers in their own learning process (Rodríguez-Inés 2009). Furthermore, the use of corpora in translation teaching offers a valuable translation toolkit. Scholars have highlighted that corpus-assisted translation teaching provides not only a reference tool but also prompts thought-provoking insights (Bernardini et al. 2003: 11).

The utilisation of different types of corpora offers distinct advantages, with monolingual corpora being particularly accessible and widely utilised. This type of corpus provides valuable insights into conventional language usage within specific contexts, empowering translators to produce more natural-sounding translations (Bernardini et al. 2003). By employing monolingual corpora of the target language, translators can search for potential translation equivalents, examine authentic language usage across diverse contexts, explore stylistic considerations, eliminate inappropriate word combinations or equivalents, and validate their intuitions (Bowker and Pearson 2002). Coffey (2002) highlights the usefulness of source-text monolingual corpora as a valuable resource for both translators and translation instructors. However, it is important to acknowledge the limitations of monolingual corpora. Whilst they can offer translations and usage examples within the same language, they do not directly provide information on how words or phrases are translated across languages. This means that translators relying solely on monolingual corpora may face challenges when searching for suitable equivalents in the target language. Moreover, monolingual corpora might not adequately address the cultural and contextual nuances that are crucial in translation. As a result, translators may need to consult additional resources or rely on their own cultural and linguistic knowledge to ensure accurate and culturally appropriate translations.

Comparable corpora, particularly comparable bilingual corpora comprising native texts in both the source and target languages, play a crucial role in the field of translation and translation teaching (Liu 2020). These corpora offer a range of valuable benefits. They provide translators with access to authentic language usage in both the source and target languages, thus addressing the issue of "translationese" to some extent (McEnery and Xiao 2007: 4). Furthermore, comparable corpora not only enhance translation students' comprehension of the distinct linguistic features present in both the target and source languages (Zanettin 1998), but they also provide invaluable insights into the cultural nuances and specialised knowledge associated with specific contexts (Zanettin 2001). However, it is important to recognise that the effectiveness of comparable corpora relies heavily on the careful selection of representative and authentic texts (Kenning 2010). Nonetheless, it is worth noting that, similar to monolingual corpora, comparable corpora cannot fully capture the intricate complexities involved in the translation process of transforming one language into another (McEnery and Xiao 2007).

The third type of corpus, referred to as parallel corpora, serves a specific purpose in translation (Zanettin 2002). Parallel corpora offer distinct advantages beyond the general benefits provided by corpus tools, such as serving as references for lexical items, syntactic structures, and stylistic concerns. What sets parallel corpora apart is their ability to provide both "direct" and "indirect" translation equivalents to translators (Zanettin 2002: 11). Moreover, parallel corpora consist of an extensive collection of source texts and their corresponding translations, enabling students to analyse diverse translation strategies employed by expert translators across various contexts (Pearson 2003). Notably, scholars emphasise the particular usefulness of parallel corpora in specialised translation, allowing translators to search for equivalent technical terms, unmarked sentence structures, and stylistic conventions within particular subject-specific genres (Kübler et al. 2015). Nevertheless, parallel corpora are comparatively less utilised in corpus-assisted translation teaching compared to monolingual and comparable corpora, partly due to the challenges involved in collecting high-quality parallel texts (Liu 2020).

In addition to comparable and parallel corpora, ad hoc corpora and learner translation corpora are also frequently employed in translation teaching by researchers and educators. Ad hoc corpora, for instance, prove to be particularly valuable in addressing specific requirements in translation (Liu 2020). Through the construction of ad hoc corpora, students engage in the selection of reliable sources (Varantola 2003) and strive to gain a deeper understanding of the meaning within the source texts (Aston and Bertaccini 2001). Another area of growing interest is the compilation of learner translation corpora, which allows for the examination of common characteristics in learner translations (Granger and Lefer 2020). Notable examples of learner translation corpora include the UPF learner translation corpus (Espunya 2014) and the undergraduate learner translator corpus (ULTC) (Alfuraih 2020).

In summary, the corpus approach in translation teaching promotes an inductive learning method, requiring active student engagement to ensure its effectiveness. It is essential to grasp the advantages and challenges of utilising corpora in translation from the students' perspective to optimise the learning process.

2.2 Using Corpora in Translation Teaching: Issues to Consider

Whilst many scholars have emphasised the benefits of using corpora in translation and have introduced various pedagogical designs for corpus-assisted translation teaching (Monzó Nebot 2008; Rodríguez-Inés 2009, 2011; Zanettin 1998, 2001, 2002), there is a relative dearth of empirical studies exploring students' performances and perceptions regarding the use of corpora in translation or translation learning. Amongst the few existing studies, Zhu and Wang (2011) developed ClinkNotes, a corpus-based tool for students' self-directed translation learning. In another study, Liu (2020) compared students' performance using parallel corpora and paper-based dictionaries in translation tasks. The findings revealed that the utilisation of parallel corpora significantly improved students' translation performance in both English-Chinese and Chinese-English translation tasks, with students also expressing a positive view of using parallel corpora in translation. However, with the growing importance of technological competence in the digital age (PACTE 2003), contemporary translators may increasingly rely on web-based resources rather than traditional paper-based dictionaries. Therefore, it is crucial to investigate the impact of parallel corpora on students' translation in more authentic and valid settings.

Despite the perceived benefits of using corpora in translation teaching, the cost-efficiency of this approach has been questioned by some researchers (Varantola 2003). Incorporating corpora into translation practice can be time-consuming as it requires training to effectively utilise corpora to meet translation needs. Moreover, successful corpus use in translation demands students' ability to critically analyse corpus data and extract relevant information from it (Bernardini 2016). The relatively low cost-efficiency of using corpora can be attributed, in part, to the limited availability of corpora specifically designed for translation teaching purposes. Accessing parallel texts, in particular, is more challenging compared to monolingual or comparable texts, resulting in the construction of small-scale parallel corpora (Zanettin 2002). In addition, many translation educators rely on existing parallel corpora primarily designed for research purposes in their teaching practices (Marco and Van Lawick 2009; Ruiz Yepes 2011). Given the challenges associated with using corpus tools in translation, there is a pressing need to design user-friendly corpora that resemble the tools familiar to translators. As Aston (2009) highlighted, the critical issue is to create corpora that can enhance translators' consultation efficiency without compromising the quality of the tool. Concerns have also been raised about students becoming overly reliant on corpus tools and potentially sacrificing the creativity of their translation output. This concern is particularly relevant for parallel corpora, which provide translation equivalents. Therefore, researchers have cautioned that corpora should not be blindly followed as absolute authorities in translation training (Bernardini et al. 2003; Malmkjær 2003).

In summary, the practical challenges surrounding the use of corpora in translation teaching and the existing gaps in empirical evidence on student performance and perceptions highlight the importance of further research in this field. These investigations will yield valuable insights for the future design and implementation of corpora in translation teaching.

2.3 Rationale and Research Questions

As highlighted in the previous review, the parallel corpus is a valuable resource in corpus-assisted translation teaching. However, its potential remains largely unexplored for various reasons. With the advancements in technology and the prevalence of translation between Chinese and English, two major languages worldwide, it is imperative to further investigate the benefits of parallel corpora. To bridge this research gap, our study specifically focuses on examining the proactive role students must assume and the potential challenges they may encounter when utilising parallel corpora in translation tasks. Through gathering empirical evidence on students' performance and perception of using parallel corpora in translation, our study aims to address two key research questions:

- How does the use of parallel corpora enhance students' translation performance?
- What are the potential challenges that students may face when utilising parallel corpora?

3 Methods

3.1 Participants

A total of 38 students voluntarily participated in this study, all of whom were enrolled in an MA translation programme at a university in Hong Kong. Through random assignment, the students were divided into two groups: an experimental group consisting of 16 students and a control group consisting of 22 students. All participants were considered intermediate-advanced English learners, with IELTS scores ranging from 6.5 to 8. In addition, over 70% of the participants from both groups reported having no prior experience using corpora in translation or English learning.

Prior to the study, informed consent was obtained from both groups of students, and the control group received remedial training. To ensure a comprehensive analysis, four students from the experimental group were purposefully selected for follow-up interviews using the principle of maximum variation. The selection process took into

Table 1 Personal profiles of the focal participants

Name	Gender	Prior practical translation experience	Prior corpus use experience	Frequency of TR corpus use
Syuki	Female	Almost no experience in translation	Occasionally used BNC in translation	Average amongst the participants
Ume	Female	Limited experience in translation	No experience	Above-average amongst the participants
Yuzi	Female	Some experience in translation	Sometimes used COCA in learning English or translation	Below-average amongst the participants
Haru	Female	Rich experience in translation	No experience	Average amongst the participants

account factors such as their previous translation experience, corpus use experience, and engagement with the parallel corpus during the study. Through an examination of divergent cases, the researcher aimed to identify common effects of the parallel corpus on different types of students and explore the potential factors that influenced students' performance and perception of corpus use in translation. Table 1 provides an overview of the selected students' profiles, with pseudonyms used to protect their identities.

3.2 The Parallel Corpus Used in the Study

The parallel corpus utilised in this study was TR Corpus (http://www.tr-corpus.com), a web-based translator training corpus specifically designed for teaching purposes. TR Corpus is a large-scale corpus constructed by sampling and compiling high-quality bilingual texts from various bilingual websites. It consists of approximately 79.31 million English words and 171.44 million Chinese characters. One of the key strengths of TR Corpus is its wide range of text types, including news articles, annual reports, company profiles, features, financial documents, and legal documents, sourced from mainland China and Hong Kong. TR Corpus offers several distinctive features in its search function, results display, and interface design, all aimed at facilitating corpus use for translation teaching and learning. The corpus provides three major functions: Search, Collocate, and Compare. These functions enable students to search for occurrences and collocations of specific words, as well as compare the meanings and usage of two words. In addition, TR Corpus includes a built-in Translator's Workbench feature, allowing students to upload parallel texts for homework submission or future review. The design and functionality of TR Corpus make it a valuable resource for students engaging in translation tasks, providing them with a comprehensive platform for searching, analysing, and comparing bilingual texts.

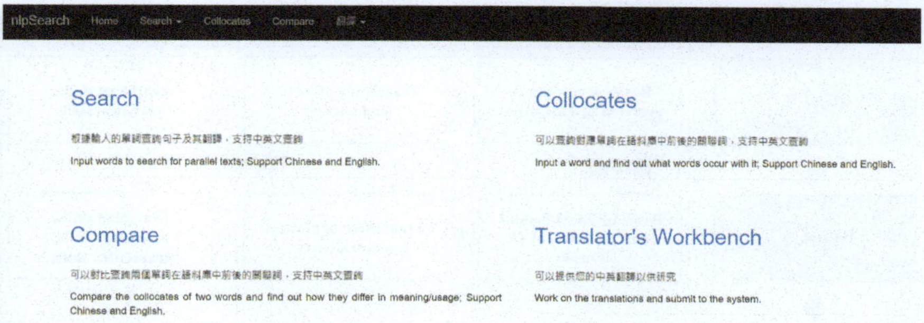

Fig. 1 Interface of TR Corpus

The search results in TR Corpus are displayed using a ranked searching mechanism, ensuring that the most relevant results appear at the top of the list. This feature significantly enhances students' translation efficiency by presenting them with the most pertinent information first. Furthermore, the search term(s) are highlighted in the search results, allowing students to quickly locate the specific instances they are interested in. To provide students with additional context, TR Corpus allows them to access the source websites of each parallel concordance. By clicking on the external link provided in the display page, students can directly visit the original source websites to obtain further information or gain a deeper understanding of the texts. In terms of interface design, TR Corpus offers a user-friendly interface with a clear navigation bar located at the top of the web page. This navigation bar enables easy access to various features and functionalities of the corpus. Below the navigation bar, there are 2 × 2 function columns, each column providing a brief introduction to the respective function it represents. This layout helps users quickly understand and familiarise themselves with the different components and functions of TR Corpus. For a visual representation of the interface design of TR Corpus, please refer to Fig. 1, which illustrates the layout and components of the corpus interface.

3.3 Procedure

A pre-test was conducted prior to the training to compare the translation performance of the experimental group and control group. The pre-test included an English-Chinese and a Chinese-English translation task, both of which involved short extracts from a company profile. During the pre-test, the students had the freedom to utilise any resources available to them.

Following the pre-test, the experimental group participated in weekly 90-min training sessions for four weeks, alongside their regular translation courses, to familiarise themselves with the use of TR Corpus in translation. In these sessions, the teacher introduced the fundamental concepts and functions of the parallel corpus.

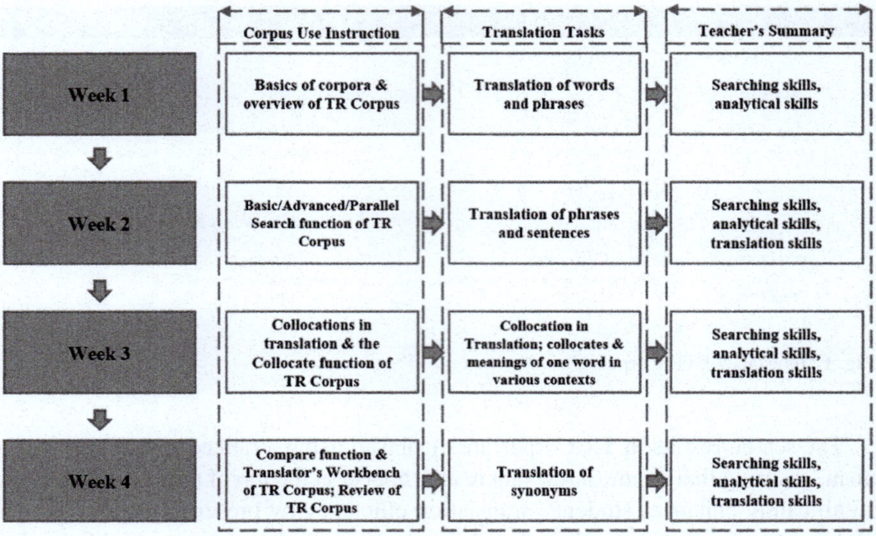

Fig. 2 Four-week parallel corpus training

Students were given dedicated time to explore the corpus, addressing linguistic issues and translation problems, and testing their intuitions using the parallel corpus. Once students became acquainted with the corpus functions, they were assigned translation tasks that encompassed various text types found in TR Corpus. These tasks aimed to encourage students' critical analysis of corpus data and the summarisation of relevant information. Students shared their findings with both their peers and the teacher. At the end of each training session, the teacher, assuming the role of a facilitator rather than an instructor, guided students in consolidating their search and data analysis skills, along with translation strategies relevant to the assigned tasks. The structure of the training sessions is outlined in Fig. 2. Meanwhile, the control group continued with their regular courses without any additional training.

Following the training, both groups of students underwent a post-test using the same text type, namely a company profile. In the post-test, the experimental group had access to the parallel corpus and specified dictionaries without machine translation functions, whilst the control group could utilise various online resources, reflecting typical translation scenarios. The majority of students completed the test within a two-hour timeframe. Subsequently, semi-structured interviews were conducted with the four focal students to gather their perceptions regarding the use of the parallel corpus in translation tasks. During the interviews, the participants were shown screencasts of their respective translation process and search history to aid their recollection and evaluation of their experiences with corpus use in translation. Each interview with a focal participant lasted approximately 40 min.

3.4 Data Collection and Analysis

The primary sources of data collected for this study comprised the pre-test and post-test translation products, transcriptions of interviews, and students' corpus search history. In addition, screencasts of the post-test translation process were gathered to provide further insights into the students' translation performance and their perceptions of utilising the parallel corpus in translation tasks.

The translation products of the students from both the pre-test and post-test were evaluated using a ten-point rating scheme adapted from Kiraly (1995: 83). To assess whether there were differences in translation performance between the experimental and control groups, an independent samples t-test was conducted to compare the pre-test results. Subsequently, to investigate the potential positive effects of using the parallel corpus on students' translation performance, another independent samples *t*-test was employed to compare the translation results of the post-test between the two groups. In addition, a quantitative textual analysis of the students' post-test translation products was conducted to further examine the disparities in translation quality between the experimental and control groups. This analysis incorporated common lexical and syntactic complexity measures frequently employed in translation studies, providing straightforward indicators of the translation strategies employed by students. The analysis involved some lexical and syntactic measures, including type/token ratio (TTR), number of sentences, and the average sentence length. However, since these simple measures may not fully capture the nuances of the translation products between the two groups, a qualitative analysis of the students' translation products was conducted. This qualitative analysis was supplemented by examining their search histories and screencasts. By combining quantitative and qualitative methods, the study aims to gain a more comprehensive understanding of the impact of using a parallel corpus on students' translation performance, as well as to discern the differences between the experimental and control groups.

All audio recordings of the interviews were transcribed verbatim. These transcriptions were then subjected to a typological analysis using three measures: usefulness, challenges, and suggestions. By conducting both quantitative and qualitative data analysis, the study aimed to infer and discuss the effectiveness and difficulties of using a parallel corpus in translation teaching.

4 Findings

4.1 Students' Translation Performances

4.1.1 Independent Samples *T*-Test

To assess the impact of utilising a parallel corpus on students' translation performance, the researchers initially conducted independent samples *t*-tests to compare the translation performance of the experimental group and the control group in the pre-test. The analysis encompassed both the Chinese-English translation task and the English-Chinese translation task. The results revealed no significant differences between the two groups in either task (Chinese-English translation: $p = 0.873$, English-Chinese translation: $p = 0.574$), indicating that both groups possessed similar translation competence prior to the experiment.

Subsequently, additional independent samples *t*-tests were conducted to compare the translation performance of the group using the parallel corpus with that of the group using regular consultation resources. In the Chinese-English translation task, although the mean score of the experimental group ($M = 7.19$, SD = 0.75) was slightly higher than that of the control group ($M = 6.77$, SD = 0.92), no significant differences were found between the two groups ($p = 0.397$). These findings suggest that the use of the parallel corpus did not have a notable impact on students' Chinese-English translation performance. Conversely, in the English-Chinese translation task, the mean score of the experimental group was 7.31 (SD = 1.08), whereas the mean score of the control group was 7.05 (SD = 0.65). Notably, the mean score of the experimental group was significantly higher than that of the control group ($p = 0.011$). This implies that employing the parallel corpus had a more beneficial effect on translation into the students' native language compared to translation out of the native language.

4.1.2 Analysis of Translation Products

To further investigate the differences in students' translation output, a textual analysis was carried out on their post-test translation products.

Chinese-English Translation

Table 2 presents the descriptive statistics of the Chinese-English translation products from the two groups. The experimental group had a slightly lower average number of tokens compared to the control group. However, their type-token ratios were similar, indicating that both groups exhibited comparable lexical variety. Although the experimental group used fewer types of words on average than the control group, individual students within the experimental group demonstrated a more diverse word choice compared to the control group. For instance, when translating the phrase "秉承…精神" (literally uphold the spirit…), students in the experimental group showed a greater variety of word choices, such as "adhere to the spirit of…" (5 students), "uphold the spirit of…" (5 students), "…in the spirit of…" (2 students), and "with the spirit of…" (2 students). In contrast, most students in the control group translated it as "adhere to the spirit/principle of…" (16 students).

In order to gain a deeper understanding of the factors contributing to the consistent use of the phrase "adhere to the spirit" in the control group, an examination of the screencasts was conducted to provide additional insights. It was observed that a significant number of students in the control group heavily relied on machine translation systems, including the translation function of online dictionaries or platforms like Google Translate. These systems consistently generated the translation "adhere to the spirit" for the given phrase, leading to a lack of variation in the translation choices amongst the students in this group.

In contrast, students in the experimental group utilised the Basic search function to search for the keyword "秉承" (*Bingcheng*, literally: uphold), as depicted in Fig. 3; or the Advanced search function to search for the keywords "秉承…精神" (*Bingcheng…jingshen*, literally: uphold the spirit…), as shown in Fig. 4. Through these searches, TR Corpus provided them with multiple translations of the phrase in various contexts. As a result, students were able to select different translations based on their own judgement or specific needs, and in the process, they gained a deeper understanding of the corresponding sentence structures through the examples provided by the corpus.

Figure 4 reveals that the control group exhibited a wider variety of lexical terms and produced a greater number of words. In contrast, the experimental group employed a "splitting" translation strategy, breaking down lengthy Chinese sentences into shorter ones, resulting in shorter average sentence lengths. This strategy, learned from analysing the parallel corpus data, aimed to ensure natural-sounding translations. Notably, a student from the experimental group mentioned during an interview that the corpus results influenced her adoption of the "splitting" strategy.

> Initially, I was unsure how to handle the sentence's length. However, upon searching for "發揮優勢" (*Fahui youshi*, literally develop advantages) in the company profile subcorpus, I discovered numerous similar lengthy sentences. This experience helped me realise the need to divide this lengthy sentence into shorter ones.

Table 2 Descriptive Statistics of Chinese-English translation

Description	Experimental group (mean)	Control group (mean)
Tokens	174.31	175.82
Types	96.69	99.55
Type/token ratio (TTR)	0.54	0.55
Sentences	5.81	5.45
Mean sentence length (in words)	31.59	34.11

This was especially the case at Crystal Palace, where a 1990 act of parliament specified that any new building on the site must be in the spirit of the original. (url)	水晶宫项目尤其如此，1990年的一项议会法案明确提出，任何在原址上的新建筑都必须秉承原物风貌。
Like much else about the current administration, Mr. Spicer's appearance breaks with tradition, in this case of customs long maintained by those who serve as the president's official mouthpiece. (url)	像当前政府中的其他很多人一样，斯派塞的外表打破了传统，在此是指充当总统的官方喉舌的那些人长期秉承的传统。
The abbot acknowledged that it might seem strange for Taoists, who practice a philosophy of wu wei, or inaction, to be leading a call for change. Still, he said it was important to set an example. (url)	杨道长承认，由秉承无为哲学的道家来领导一项呼吁改变的行动，看起来可能是有点奇怪。但他表示树立一个榜样是很重要的。
"His conduct went against the values of 'equality and respect' that Daimler has long upheld," Daimler said in a statement. "We have immediately relieved the manager of his position," it said. (url)	戴姆勒在声明中称："他的行为与戴姆勒所秉承的'平等、尊重'的价值观相违背。我司立即对该经理人进行了免职处理。"
Southern Yangtze University adheres to its Motto of"Diligent in learning and committed to doing till you are a perfect being"and with its centennial achievements of several distinguishing disciplines, is now striving for even greater success in the new century to become a high level comprehensive university of global reputation. (url)	学校秉承"笃学尚行，止于至善"的校训，为在新的世纪里建设成世界知名的高水平综合性大学而奋斗。
The company said in an emailed statement that it "has long been adhering to the principle of lawfulness and compliance in all our business operation and international exchanges and will carry on following such a principle". (url)	该集团在一份电子邮件声明中称，"我们在所有业务运作和国际交流中一直秉承了合法、合规的原则，并将继续遵循这一原则。"

Fig. 3 Screenshot of Basic Search Results for the Keyword "秉承 (*Bingcheng*)" in TR Corpus

But bear with me. More in the spirit of Twitter than column-writing – as in, let's toss out some thoughts and see how the crowd reacts – here are a few other factors that may make the internet a hospitable place for malice: (url)	请对我耐心一点，让我秉承着写Twitter而非写专栏的精神——即随便抛出一些想法，看看大众如何反应——讲一讲几个让互联网变成藏污纳垢场所的其他因素：
The UNESCO Medal on Space Science, established on June 29, 2017 is awarded by the UNESCO director-general to honor prominent scientists, public figures and organizations for their contributions to the development of space science in the spirit of UNESCO's priorities. (url)	联合国教科文空间科学奖章于2017年6月29日建立。此奖章由联合国教科文组织总干事授予对秉承联合国科教文组织精神，并且对空间科学的发展作出重要贡献的科学家，公众人物和组织。
On the talk show A Date With Luyu, which was shot just a dozen days before she passed away, Xiong laughed as she shared her experience in the hospital, saying she had only cried once over her illness. Her optimism and courage have inspired millions of people, and the film is consistent with her spirit. (url)	去世十多天前，熊顿参加了《鲁豫有约》的录制。她在节目中笑谈自己住院期间的故事，还坦言自己只为肿瘤哭过一次。她的乐观和勇气激励了成千上万的人，电影则秉承了她的这一精神。
In much the same spirit of seeking new talent, LVMH is continuing to narrow its candidate list for the first LVMH Prize, to be awarded in May. (url)	路威酩轩集团秉承同样的寻找新人才的精神，继续精简它的第一个路威铭轩奖的候选人名单，该奖项将在5月份颁发。
He sought to control his workforce, too. In 1976, he wrote a manifesto, "The Testament of a Furniture Dealer," with biblical-style commandments listing simplicity as a virtue and waste as a sin. Employees were expected to absorb "the Ikea spirit," to be humble, clean-cut and courteous, not just knowledgeable about Ikea's products but enthusiastic about its corporate ideology — principles to work and live by. (url)	他还试图控制自己的员工。1976年，他写了一份名为《一个家具经销商的遗嘱》(The Testament of a Furniture Dealer)的宣言，列出了圣经式的戒条，称简朴是美德，浪费是罪恶。他希望员工秉承"宜家精神"，做到谦逊、整洁、有礼，不仅对宜家的产品了如指掌，还要对公司的理念，即工作和生活原则充满热情。
The new recruits are redefining the design programs of those institutions, notably by analyzing design within a social and political context, rather than from a traditional aesthetic perspective. The V&A has radicalized its approach to design collecting since Mr. Long's arrival, and Mr. Chen is assembling a collection of Asian design from scratch at M+ in the same spirit. (url)	这些策展人正在新职位上重新定义这些机构的设计项目，主要是在社会和政治环境中分析设计，而不是从传统的审美角度。基兰·朗加入维多利亚与艾伯特博物馆之后，彻底改变了设计收藏方法。陈伯康秉承同样的精神，在M+博物馆从头收集亚洲设计藏品。
Mr. Xi, who was appointed Communist Party chief in November and is set to become state president in March, has said he wants to accelerate economic changes in the spirit of Deng Xiaoping, who began the process of transforming China into a more modern economy after decades of rule by Mao Zedong. But in the process of introducing market forces in China, such changes have starkly widened income disparities. (url)	11月就任中国共产党总书记的习近平将在3月成为国家主席。他曾表示，要秉承邓小平的精神，加快经济变革。在毛泽东数十年的统治结束之后，邓小平开启了将中国转变成一个更加现代化的经济体的过程。但是在引入市场力量的过程中，这些变革明显扩大了中国的收入差距。

Fig. 4 Screenshot of Advanced Search Results for the Keyword "秉承…精神(*Bingcheng…jingshen*)" in TR Corpus

Example 1 In the target text, one sentence from the source material has been divided into three distinct sentences.

Source Text:

科大國創源自中國科學技術大學，擁有一支高水平的研發團隊，秉承"務實、創新"的精神，肩負"軟件興企報國，創新引領未來"的偉大使命，發揮多年積累的軟件與大數據技術和深厚的行業經驗優勢，抓住人工智能發展契機，積極開展數據智能技術的研發和應用，構建領先的數據智能核心技術，賦能各行業領域客戶專屬的數據智能能力，推動國家以數據為驅動的數智化轉型。

Translation Example from the Experimental Group (5 sentences):

We are originally from The University of Science and Technology of China and have thus benefitted with a high-level R&D team. // In the spirit of pragmatism and innovation, our mission is to reward the country via future innovation in software excellence. // We make the best of our great experience in the industry and our technology in software and big data amid the golden era for AI development. // We not only proactively engage in R&D and the application of digital intelligence, but also pioneer the development of cutting-edge core technology for digital intelligence. // By doing so, we strive to assist customers across industries with digital intelligence capacities and promote digital intelligence transformation, driven by big data.

Translation Example from the Control Group (2 sentences):

Guochuang Software originated from the University of Science and Technology of China, with a high-level R&D team, adhering to the spirit of "pragmatism and innovation", shouldering the great mission of "software for the enterprise to serve the country, and innovation leading the future". // With big data technology and profound industry experience advantages, the company catches up the opportunity of artificial intelligence development, actively carries out research and development and the application of data intelligence technology, builds up leading data intelligence core technology, empowers customers in various industries and fields and promotes the country in the transformation to digital intelligence driven by data usage.

English-Chinese Translation

Table 3 presents the descriptive statistics of the English-Chinese translation products from the two groups. The control group demonstrated a higher average use of characters and words in their English-Chinese translation compared to the experimental group. However, the experimental group exhibited a relatively higher Type-Token Ratio (TTR), indicating greater lexical variation in their translation products. Upon comparing the English-Chinese translation products of the two groups, it was observed that some students in the experimental group employed an "omission" strategy to enhance text cohesion. For instance, in Example 2, a student from the experimental group used this strategy to translate the proper noun "In-Tech" in the source text. The student rendered it as "誠科" (*Chengke*, literally "In-Tech") in the first sentence, omitted it in the second sentence, and translated it as "我們" (*Women*, literally "We") in the third sentence. In contrast, a student from the control group strictly adhered to the source text's sentence structure and translated three sentences with an identical subject "本公司" (*Ben gongsi*, literally "Our company").

Table 3 Descriptive statistics of English-Chinese translation

Description	Experimental group (mean)	Control group (mean)
Characters	418.38	434.95
Tokens in text	227.63	234.14
Types	144.56	145.52
Type/token ratio (TTR)	0.64	0.62
Sentences	12.8	12.71
Mean sentence length (in characters)	33.7	33.75

Example 2
Source Text:

In-Tech offers turnkey solutions for new projects, as well as supplying electronic assemblies and completed products. In addition, **In-Tech** also uses its workshop in Hong Kong to provide quick turn repairs, refurbishment and order fulfilment services. **In-Tech**'s quality management is accredited to serve aerospace, automotive and medical customers.

Translation Example from the Experimental Group:

誠科不僅為新項目提供一站式解決方案及電子產品，亦透過香港廠房提供產品維修、翻新及訂單履行服務。我們針對航天、汽車及醫療行業的質量管理已獲得相關認證。

(Back translation in English: **Chengke** not only provides one-stop solutions and electronic products for new projects, but also provides product repair, refurbishment and order fulfillment services through its Hong Kong factory. **We** have obtained relevant certifications for our quality management in the aerospace, automotive and medical industries.)

Translation Example from the Control Group:

本公司为新项目提供一站式解决方案，并提供电子组件和成品。此外，本公司亦利用其在香港的工作坊，提供快速维修、翻新及订单履行服务。本公司质量管理已经认可，可为航空航天、汽车和医疗客户服务。

(Back translation in English: **Our company** provides one-stop solutions for new projects and provides electronic components and finished products. In addition, **our company** also uses its workshops in Hong Kong to provide rapid repair, refurbishment and order fulfilment services. **Our company's** quality management is accredited and serves aerospace, automotive and medical customers.)

Both the experimental group and the control group exhibited a similar average number of sentences in their English-Chinese translations, indicating that students from both groups made intentional adjustments to the sentence structures. As a result, the translation outcomes were rather comparable between the two groups.

4.2 Perceptions of Students

In general, students expressed a positive attitude towards utilising the parallel corpus in their translation work, regardless of their varying levels of experience in translation and corpus use. However, they also acknowledged encountering certain challenges during the process. The students also provided valuable suggestions for enhancing the design of the parallel corpus, aiming to further improve its effectiveness and usability in translation practice.

4.2.1 Advantages of Using the Parallel Corpus in Translation

Ease of Use and Reliability

One prominent advantage of the parallel corpus, as highlighted by the participants, is its user-friendly design. Syuki, in particular, praised the ease of use of the TR Corpus, noting its suitability for students with limited experience in utilising translation technology tools. She expressed her appreciation for the user-friendliness of the parallel corpus to such an extent that she voiced concerns about potential challenges if the corpus functions were to become more complex in the future.

> I'm sure TR Corpus will get better and add more parallel data in the future. I hope it becomes more professional, but at the same time, I'm a bit worried that it might become too complicated for me to use.

Yuzi, who had some proficiency in using corpora, also agreed that the corpus was user-friendly and mentioned that she might be able to use it without additional training. All four participants acknowledged that the corpus results were more reliable compared to other online resources. Ume specifically mentioned that the corpus data were more trustworthy than search engines like Baidu or Google, which could potentially provide numerous low-quality translations.

> If you're using a search engine like Google, you might come across translations uploaded by unknown netizens… But with TR Corpus, the results are more reliable, especially when it comes to professional terms.

Haru compared the reliability of corpus results with Google Translate, noting that machine translation can often misinterpret the meaning or context of the source text, resulting in inaccurate word-by-word translations. In contrast, when she searched for something on TR Corpus, she found highly reliable references accompanied by abundant examples in various contexts.

Providing Translation References

The participants generally held a positive attitude towards the extensive collection of translation references provided by the parallel corpus. Depending on their prior experience with corpus use, they employed translation equivalents to varying degrees. Ume, in particular, who lacked confidence in her translation abilities, heavily relied on translation equivalents extracted from the parallel corpus. By examining multiple versions of translation equivalents, she could compare their usage in different contexts and select the most suitable one, leading to a successful performance in the post-test. Furthermore, Ume emphasised the significant role played by the parallel corpus in addressing her long-standing concerns regarding collocation in translation. Through the corpus, she discovered valuable solutions to her collocation issues, resulting in significant improvements in the quality of her translations.

Despite Syuki's prior experience using the BNC in translation, she did not specifically refine her search strings to extract direct translation equivalents from the corpus. Instead, she discovered that analysing the language use in diverse contexts provided her with a deeper understanding of the meaning and usage of words or phrases in the source texts. This approach proved beneficial in generating more appropriate translations. In contrast, Haru shared that her approach to using the corpus varied depending on the type of text she was translating. For familiar texts such as news or company introductions, her focus was on comprehending the meaning and usage of phrases. However, when tackling legal translations with distinct lexical, syntactic, and stylistic features, she shifted her attention to sentence patterns that might not be readily accessible through conventional consultation resources.

In contrast to the other three students who found the Search function valuable for obtaining reference translations, Yuxi had a preference for using the Compare and Collocate functions of TR Corpus to explore translation equivalents. When faced with uncertainty about which word was more suitable for translation, she would compare the meanings and usage of two words using the Compare function. In addition, Yuxi frequently relied on the Collocates function to search for collocations, aiming to ensure that her translations sounded more natural in the target language.

Besides utilising the parallel corpus to address lexical translation challenges, the students also highlighted the advantages of using it to tackle textual or stylistic issues. Ume, for example, mentioned that,

> Sometimes, when I'm not sure about my translation, I search for keywords and find a bunch of sentences as references. I learn from the sentence patterns in the examples to make sure that my translation style is appropriate.

Yuzi also emphasised how the corpus helped her become more familiar with different text types. In her own words:

> When I searched for keywords, I came across numerous parallel texts (of the same text type). I would click on the links to read the source websites and get a better grasp of the text type.

All four participants agreed that the parallel corpus was especially valuable for translating specialised text types, particularly in the field of legal translation. Syuki specifically noted that "certain industry jargon may be challenging to locate through other means".

Improving Translation Efficiency

All four participants acknowledged that the corpus yielded more reliable results compared to search engines such as Baidu or Google. This advantage of the corpus design further enhanced translation efficiency. Syuki and Yuzi attributed their increased efficiency to the trustworthy nature of the parallel corpus data, as they no longer needed to spend time verifying the credibility of the data sources. Haru, who was adept at utilising various translation search techniques, noted that she could avoid getting overwhelmed by excessive data and instead quickly identify the relevant information from the corpus results.

Enhancing Translation Confidence

All four participants expressed that the corpus had contributed to an increased sense of confidence in their translation abilities. Syuki and Ume, who had less experience and confidence, appreciated the opportunity to learn and utilise the new tool in their translation work. In addition to acquiring new knowledge and skills, their growing confidence could be attributed to the parallel corpus serving as a means to validate and confirm their translation choices. Ume specifically mentioned feeling more assured when her translation intuitions aligned with the corpus data. Syuki also highlighted the corpus's role in boosting her confidence, particularly in the domain of Chinese-English translation:

> It's more challenging for me to do Chinese-English translation. I didn't believe in myself, but I trust TR Corpus. With TR Corpus, I can look at the translation examples done by expert translators.

Yuzi and Haru, who have greater translation experience compared to the other participants, expressed a similar viewpoint regarding the important role of the parallel corpus in validating their translation intuitions. In situations where uncertainties or gaps in their memory arise during the translation process, both Yuzi and Haru turn to the parallel corpus to validate their understanding and improve the accuracy of their translations. Yuzi specifically highlighted the role of the parallel corpus as a dependable resource when she finds herself "unsure of her memory".

4.2.2 Challenges and Suggestions

Limitation of Corpus Design

Although the parallel corpus offers a vast amount of data and operates in a user-friendly manner, students acknowledged that there were instances when they couldn't locate the desired information within TR Corpus. The primary reason cited for this limitation was the restricted availability of text types in the corpus. With only six text types currently included, students found it less advantageous when translating texts outside of those categories, such as literary translations. In addition, occasional server capacity issues caused the corpus to fail in loading results, particularly during peak usage periods when the entire class attempted to access it simultaneously. As a result, students experienced delays and reduced performance as the corpus response time slowed down.

During the interview, Haru expressed her frustration with the intermittent connectivity and lag issues she encountered whilst using the platform for searching. She speculated that it could be due to her usage of incorrect search strings, but regardless of the cause, she found it exasperating to repeatedly face this problem:

> The platform kept disconnecting and lagging continuously while I was searching. I'm not sure if it's because I used the wrong search terms, but it really frustrates me when I come across this issue multiple times.

Ume also got frustrated when she couldn't retrieve the corpus results she needed. On the other hand, Yuzi opted to rely on alternative tools for assistance when the corpus failed to load or provided unusual outcomes.

Apart from the technical issues with the corpus, participants occasionally encountered difficulties in finding translation equivalents for specific keywords and found it time-consuming to analyse corpus examples with excessively long sentences. This challenge could be attributed to the design of the parallel corpus, particularly in terms of text segmentation and alignment. As Syuki pointed out:

> It's like, sometimes the alignment of sentences in the corpus doesn't really match up, you know? In real translation work, we often have to reconstruct the text using different translation strategies. So maybe instead of aligning the texts based on sentence structure, they could align them based on the meaning, you know what I mean?

Inadequate Search Skills

One challenge that emerged was the students' insufficient search skills when using the corpus. In particular, the messy results they encountered can be attributed, at least in part, to their inappropriate selection of search words. The focal participants primarily relied on the corpus to find lexical references that would assist them in their translations. Although the teacher emphasised the importance of selecting appropriate search words to maximise the corpus's effectiveness, the students appeared to struggle in this aspect. For instance, Syuki, who had less experience, faced difficulties in identifying sentence patterns from the examples in her own translation work. Consequently, she primarily utilised the corpus to access the meanings and usage of specific lexical items. In contrast, Ume demonstrated greater proficiency in adapting search string combinations to locate desired translation equivalents and sentence patterns. Notably, Ume paid attention to sentence patterns alongside search keywords and phrases in the parallel translation occurrences.

Lack of Critical Analysis

Apart from the challenge of lacking effective search skills, students also faced difficulties in critically analysing the corpus data. Whilst they acknowledged the value of the parallel corpus in providing direct reference translations that demonstrate how certain terms and expressions are translated in context, they struggled when asked about their selection criteria for specific translation versions. The students often relied on high-frequency translation versions in the corpus or simply chose a single translation equivalent for their search keywords without engaging in deeper critical analysis. Through interviews and screencasts that captured their decision-making processes, it became evident that further training in the critical evaluation of corpus translation examples is necessary to enhance students' translation awareness and foster their critical thinking skills.

5 Discussion

This study employed an experimental design to investigate students' performances and perceptions of using the parallel corpus in translation tasks. The results of the independent samples t-test revealed that the use of the parallel corpus did not have a significant impact on students' translation performance in Chinese-English translation, but it did in English-Chinese translation. These findings differ from those of Liu (2020), who reported that the use of a parallel corpus significantly improved students' translation performance in both English-Chinese and Chinese-English translation.

The disparities in the findings may be attributed to differences in the research designs employed in the two studies and variations in the English proficiency levels of the students. In Liu's (2020) study, the control group was limited to using paper-based dictionaries, whereas in the present study, the control group had access to various online and offline consultation resources except for TR Corpus. It is important to note that the experimental group in this study could only access TR Corpus and designated dictionaries. This discrepancy in resource availability between the two groups might explain the relatively similar performance observed in Chinese-English translation tasks, as the experimental group was restricted from consulting other online resources such as machine translation tools and search engines. However, in the English-Chinese translation tasks, the experimental group achieved significantly higher scores than the control group. This discrepancy in performance suggests that the translation direction could be an influential variable affecting the effectiveness of corpus use in translation tasks.

Although the statistical analysis of the translation post-test did not reveal significant differences between the two groups in indicators such as type-token ratio and sentence length, a qualitative analysis of the translation products highlighted substantial variations in both tasks. Specifically, in Chinese-English translation, students in the experimental group exhibited more distinctive word choices compared to the control group. This finding contradicts the notion that corpus use may promote conservatism in translation and impede students' creativity, as suggested by Malmkjær (2003). The availability of a parallel corpus provides students with a range of translation equivalents and reference translations, thereby offering them a broader array of choices that can be selected based on their understanding of the context. Furthermore, the analysis of the translation products from both translation tasks revealed that the experimental group employed diverse translation strategies, including sentence splitting and omission. This observation supports the perceived advantages of using a parallel corpus in enabling student translators to acquire translation strategies from the work of professional translators (Pearson 2003).

The use of a parallel corpus has proven effective in enhancing students' translation skills, particularly in terms of word choice and the acquisition of translation strategies. Interviews conducted as part of this study further validate the benefits of incorporating a parallel corpus into translation teaching and learning. Previous research based on surveys has highlighted the challenges students face when learning to use corpus tools (Zhu and Wang 2011). However, the students in our study expressed appreciation for the user-friendly nature of the corpus platform, which plays a crucial role in influencing their willingness to adopt the tool in their learning process (Charles 2014). The corpus design also takes into account the cost-efficiency issues associated with learning corpus tools, as noted by Varantola (2003).

Furthermore, our study indicates that students' prior knowledge and experience influence their utilisation of the corpus tool. Despite their individual focuses, all participants in the experimental group recognised the benefits of the parallel corpus in addressing lexical translation challenges. This finding aligns with the results reported by Liu (2020), who found that the parallel corpus is more effective in resolving micro language issues rather than macro ones in translation. Whilst the experimental results demonstrated that the parallel corpus was more effective in assisting students with English-Chinese translation compared to the reverse direction, students' perceptions expressed during the interviews were mixed. The influence of translation direction as a significant variable (Campbell 1998) on the efficacy of the corpus becomes evident. The students' favourable assessment of the parallel corpus in Chinese-English translation can be attributed to their reliance on the corpus for support when translating into a foreign language. Furthermore, students' perceptions may be directly shaped by the corpus design, which comprises a greater number of Chinese-English texts compared to English-Chinese texts. To address these considerations, future studies should incorporate translation direction as a factor in parallel corpus compilation.

Despite the overall positive attitude of students towards using the parallel corpus in translation, they also faced certain challenges. These challenges pertained to the design of the corpus and a lack of effective search and analytical skills for critically evaluating corpus data. Consequently, it is essential for teachers to offer further guidance to students regarding proficient corpus searching techniques and the critical analysis of corpus data (Bernardini 2016). In addition, considering the varying perspectives of students in this study regarding the benefits of the parallel corpus in translation, teachers can foster a collaborative learning community where students can exchange their experiences and insights on corpus usage with one another. This platform would enable students to learn from each other's approaches and enhance their understanding and utilisation of the corpus in translation tasks.

6 Conclusion

The present study aimed to investigate the advantages and challenges associated with using a parallel corpus in translation by analysing students' performance and perceptions in an experimental study. The findings from both students' performance and their perceptions indicate that, overall, the parallel corpus is considered a valuable tool in translation. Its use has resulted in increased awareness of translation problems and enhanced resourcefulness amongst students.

However, the study does have certain limitations that need to be acknowledged. Firstly, the pre-training and post-training tests were conducted at different difficulty levels, which hindered the ability to determine whether students' performance significantly improved after receiving corpus training. Secondly, the analysis of translation products only considered a limited set of lexical and syntactic indices. In future research, it would be beneficial to incorporate a wider range of indices to thoroughly assess the lexical and syntactic complexity of students' translations. In addition, it is worth noting that the corpus training in this study was conducted as an extra-curricular activity and lasted for only four weeks, which may not have been sufficient to fully equip students with proficient corpus skills. Future studies could employ a longitudinal design, integrating the parallel corpus into regular translation courses, in order to track the progress and conceptual development of students over an extended period of time.

Despite the aforementioned limitations, the current study provides practical implications for corpus-assisted translation teaching in terms of corpus compilation and pedagogical design. Firstly, with regard to corpus design, it is recommended to include a wider range of text types whilst considering the directionality of the corpus data. Moreover, to enhance the user experience of the parallel corpus, it is advisable to increase server capacity to accommodate simultaneous access by students in real teaching settings. Secondly, in corpus-assisted translation teaching, it is important to maintain a balance between translation knowledge and corpus knowledge. Placing too much emphasis on corpus skills may lead to an uncritical reliance on the corpus without proper evaluation of the appropriateness of translations in different contexts. Finally, when applying the parallel corpus to English-Chinese and Chinese-English translation, teachers should be attentive to the differences that may arise between the two directions.

References

Alfuraih, Reem F. 2020. The undergraduate learner translator corpus: A new resource for translation studies and computational linguistics. *Language Resources and Evaluation* 54 (3): 801–830.

Aston, G. 2009. Forward. In *Corpus Use in Translating: Corpus Use for Learning to Translate and Learning Corpus Use to Translate*, eds. Alison Beeby, Patricia Rodriguez-Inés, and Pilar Sánchez-Gijón, IX-X. Amsterdam: John Benjamins.

Aston, Guy and Franco Bertaccini. 2001. Going to the clochemerle: Exploring cultural connotations through ad hoc corpora. In *Learning with Corpora*, ed. Aston Guy, 198–219. Houston, TX: Athelstan.

Bernardini, Silvia. 2004. Corpus-aided language pedagogy for translator education. In *Translation in Undergraduate Degree Programmes*, ed. Kirsten Malmkjær, 97–111. Amsterdam: John Benjamins.

Bernardini, Silvia. 2016. Discovery learning in the language-for-translation classroom: Corpora as learning aids. *Cadernos De Tradução* 36: 14–35.

Bernardini, Silvia, Dominic Stewart, and Federico Zanettin. 2003. Corpora in translator education: An introduction. In *Corpora in Translator Education*, ed. Federico Zanettin, Silvia Bernardini, and Dominic Stewart, 1–13. Manchester: St. Jerome.

Biber, Douglas, Susan Conrad, and Randi Reppen. 1998. *Corpus Linguistics: Investigating Language Structure and Use*. Cambridge: Cambridge University Press.

Boulton, Alex, and Tom Cobb. 2017. Corpus use in language learning: A meta-analysis. *Language Learning* 67 (2): 348–393.

Bowker, Lynne. 2002. Working together: A collaborative approach to DIY corpora. In *Language Resources for Translation Work and Research. LREC 2002 Workshop Proceedings*, Spain, 29–32.

Bowker, Lynne, and Jennifer Pearson. 2002. *Working with Specialised Language: A Practical Guide to Using Corpora*. London: Routledge.

Campbell, Stuart. 1998. *Translation into the Second Language*. London: Longman.

Charles, Maggie. 2014. Getting the corpus habit: EAP students' long-term use of personal corpora. *English for Specific Purposes* 35: 30–40.

Coffey, Stephen. 2002. Using a source language corpus in translator training. *Intralinea Online Translation Journal* 5: 14–16.

Espunya, Anna. 2014. Investigating lexical difficulties of learners in the error-annotated UPF learner translation corpus. In *Twenty Years of Learner Corpus Research. Looking Back, Moving Ahead: Proceedings of the First Learner Corpus Research Conference (LCR 2011)*, vol. 1, ed.

Sylviane Granger, Gaëtanelle Gilquin, and Fanny Meunier, 129. Louvain-La-Neuve: Presses Universitaires de Louvain.

Frérot, Cécile. 2016. Corpora and corpus technology for translation purposes in professional and academic environments. Major achievements and new perspectives. *Cadernos De Tradução* 36: 36–61.

Granger, Sylviane, and Marie-Aude. Lefer. 2020. The multilingual student translation corpus: A resource for translation teaching and research. *Language Resources and Evaluation* 54 (4): 1183–1199.

Kenning, Marie-Madeleine. 2010. What are parallel and comparable corpora and how can we use them? In *The Routledge Handbook of Corpus Linguistics*, ed. Anne O'Keeffe and Michael McCarthy, 487–500. London: Routledge.

Kübler, Nathalie, Mojca Pecman, and Alexandra Volanschi-Mestivier. 2015. A study on the efficiency of corpus use for translation students during terminology processing and LSP translation. *CULT Conference*, 26–29.

Kiraly, Donald C. 1995. *Pathways to Translation: Pedagogy and Process*. Kent, Ohio: Kent State University Press.

Laviosa, Sara. 2008. *Description in the Translation Classroom*. Amsterdam/Philadelphia: John Benjamins.

Liu, Kanglong. 2020. *Corpus-Assisted Translation Teaching*. Singapore: Springer.

Marco, Josep, and Heike Van Lawick. 2009. Using corpora and retrieval software. In *Corpus Use and Translating: Corpus Use for Learning to Translate and Learning Corpus Use to Translate (Vol. 82)*, eds. Allison Beeby, Patricia Rodríguez Inés, and Pilar Sánchez-Gijón, 9–28. John Benjamins Publishing.

Malmkjær, Kirsten. 2003. On a pseudo-subversive use of corpora in translator training. In *Corpora in Translator Education*, ed. Federico Zanettin, Silvia Bernardini, and Dominic Stewart, 119–134. Manchester: St. Jerome.

McEnery, Tony, and Richard Xiao. 2007. Parallel and comparable corpora: What is happening? In *Incorporating Corpora*, eds. Gunilla Anderman and Margaret Rogers, 18–31. Multilingual Matters.

Monzó Nebot, Esther. 2008. Corpus-based activities in legal translator training. *The Interpreter and Translator Trainer* 2(2): 221–252.

PACTE. 2003. Building a translation competence model. In *Triangulating Translation: Perspectives in Process-Oriented Research*, ed. Fábio Alves dos Santos, 43–46. Amsterdam: John Benjamins.

Pearson, Jennifer. 2003. Using parallel texts in the translator training environment. In *Corpora in translator education*, ed. Federico Zanettin, Silvia Bernardini, and Dominic Stewart, 15–24. Manchester: St. Jerome.

Rodríguez-Inés, Patricia. 2009. Evaluating the process and not just the product when using corpora in translator education. In *Corpus Use and Translating: Corpus Use for Learning to Translate and Learning Corpus Use to Translate (Vol. 82)*, eds. Allison Beeby, Patricia Rodríguez Inés, and Pilar Sánchez-Gijón, 129–149. John Benjamins Publishing.

Rodríguez-Inés, Patricia. 2011. Electronic corpora and other information and communication technology tools: An integrated approach to translation teaching. *The Interpreter and Translator Trainer* 4 (2): 251–282.

Ruiz Yepes, Guadalupe. 2011. Parallel corpora in translator education. *Electronic Journal of Didactics of Translation and Interpretation* 7: 65–80.

Santos, Diana, and Ana Frankenberg-Garcia. 2007. The Corpus, its users and their needs: A user-oriented evaluation of COMPARA. *International Journal of Corpus Linguistics* 12 (3): 335–374.

Varantola, Krista. 2003. 5.2 Linguistic corpora (databases) and the compilation of dictionaries. In *A Practical Guide to Lexicography*, ed. Piet van Sterkenburg, 228–239. John Benjamins.

Zanettin, Federico. 1998. Bilingual comparable corpora and the training of translators. *Meta: journal des traducteurs/Meta: Translators' Journal* 43(4): 616–630.

Zanettin, Federico. 2001. Swimming in words: Corpora, translation and language learning. *In Learning with Corpora*, ed. Aston Guy, 1000–1021. Houston, TX: Athelstan.

Zanettin, Federico. 2002. Corpora in translation practice. *Third International Workshop on Language Resources for Translation Work, Research & Training, Italy*, 10–14.
Zhu, Chunshen, and Hui Wang. 2011. A corpus-based, machine-aided mode of translator training: ClinkNotes and beyond. *The Interpreter and Translator Trainer* 5 (2): 269–291.

Kanglong Liu is Assistant Professor at the Department of Chinese and Bilingual Studies of the Hong Kong Polytechnic University. He specialises in corpus-based translation studies and his main interests include empirical approaches to translation studies, translation pedagogy and corpus-based translation research. He is currently Associate Editor of Translation Quarterly, the official publication of the Hong Kong Translation Society.

Yanfang Su is currently a Ph.D. student in the Department of Chinese and Bilingual Studies, The Hong Kong Polytechnic University. Her research interests include corpus linguistics, corpus-based translation studies, and computer-assisted language learning. She has previously published in linguistic journals such as Language Learning & Technology and System and contributed book chapters on corpus-based translation studies and language learning.

Dechao Li is Professor of Department of Chinese and Bilingual Studies, The Hong Kong Polytechnic University. He also serves as the chief editor of *Translation Quarterly*, a journal published by the Hong Kong Translation society. His main research areas include corpuses-based translation studies, empirical approaches to translation process research, history of translation in the late Qing and early Republican periods and PBL and translator/interpreter training.

Learner Corpora

Data Acquisition and Other Technical Challenges in Learner Corpora and Translation Learner Corpora

Adam Obrusnik

1 Introduction

Corpora consisting of language learner texts have been compiled and utilised in linguistics research since the early 90s. Considering learner corpora, there are two fundamentally different types of resources and architectures. A learner corpus (LC) is, typically, a collection of L2 student essays or other assignments. As such, a learner corpus is effectively a monolingual corpus. A learner translation corpus (LTC) is typically a collection of source language (SL) texts and their target language (TL) translations, carried out by L2 learners. The obvious advantage of LTC over LC is the possibility to investigate SL intrusion or transfer from the mother tongue whilst the main disadvantage is more complicated data processing and querying.

Furthermore, both LC and LTC can contain an error annotation layer. Error annotations indicate the positions and types of errors that the L2 learner has made and they typically need to be entered manually by an experienced language tutor, who is following a well-defined methodology.

Whilst LC are typically leveraged in learner corpus research, LTC are typically utilised in corpus-based translation studies, as discussed by Granger and Lefer (2020). The cited publication contains various instances, where LC and LTC were successfully used and also focuses on the current challenges of the two research directions, including the lack of communication between them, lack of consistent methodology and limited re-usability of the assembled datasets. However, the aim of this publication is not to discuss the methodological and research challenges but rather, it focuses on a much more practical topic, which is typically not discussed in linguistics research papers—the technical challenges related to data acquisition, compilation and processing in LC and LTC.

A. Obrusnik (✉)
Masaryk University, Brno, Czechia
e-mail: adam.obrusnik@gmail.com

J. Pan and S. Laviosa (eds.), *Corpora and Translation Education*, New Frontiers in Translation Studies, https://doi.org/10.1007/978-981-99-6589-2_8

The aim of this chapter is to illustrate that integrating the entire data collection, data processing and data exploitation pipeline into a common web-based interface, greatly facilitates building of LC and LTC resources. In the past, this has been illustrated on the Hypal software used by the research group of Fictumova (Fictumová et al. 2017) and Hypal4MUST software co-designed and used by Granger and collaborators within the MUST project (Granger and Lefer 2020).

2 Data Acquisition

The first major challenge discussed in this chapter is data acquisition. Although it is often not explicitly mentioned in the pioneering works focusing on LC and LTC (Fictumová et al. 2017; Granger 2017; Laviosa 2012), there are several obstacles in obtaining data for LC and LTC. The first obstacle, which is discussed in this subsection, is related to obtaining the data in a consistent format.

It is widely known amongst corpus researchers, that in order to compile and query a corpus, the documents have to be provided in some sort of markup language. A markup language is a way of enriching plain text by additional information—most commonly typographic instructions.

2.1 State-of-the-Art

As mentioned in the introduction, most secondary and tertiary education institutions nowadays offer language courses or even translation courses. Since second language education is nowadays mandatory in most countries, learner data is, in theory, abundant. But practically, the L2 learner texts are being submitted to language tutors in a plethora of formats. The most common format these days is, probably, the Office OpenXML (OOXML) document format (developed by Microsoft, typically.docx file extension). The data format is quite well documented and can be transferred to a machine-readable markup language rather easily, as long as it does not contain revisions, comments or other macros (e.g. external reference manager) which the L2 learner might be using. Furthermore, even though the OOXML format is well documented, it does not guarantee perfect transferability of the document between various word processor software suites and software versions. For this reason, some tutors ask the students to submit assignments in the portable document format (PDF), which is the current industrial standard and it guarantees that the document displays identically on various computers, operating systems and software versions. However, with regard to corpus building, PDF is a rather unfortunate format—converting PDF into machine-readable markup language is difficult and oftentimes, important information is lost (e.g. locations of line breaks, font styles, bulleted lists). Annotations and comments in PDF documents are typically proprietary features of individual software suites and are always lost when trying to convert a PDF into a markup document.

2.2 Integrating Data Collection

The challenge of data collection can be overcome by integrating the data collection directly into an e-learning procedure. Practically, this means that the students are provided a web-based submission interface, where they can enter their essays or translations directly into a desktop-like text editor that they are used to.

Schematically, a data collection pipeline for L2 texts (either essays or translations), is shown in Fig. 1 whilst Fig. 2 shows the practical example of a text collection interface within the Hypal4MUST software. Apparently, the interface has to be as convenient to the students as possible, offering the SL text on the right hand side and the input window for the TL text on the left hand side. It also allows students to add most common typographic markings, which are often considered an important part of a translator's job as their incorrect usage can shift the meaning of the translation.

The advantage of this approach is that the web-based rich text editor produces a document which is already in a simple markup format (typically HTML) which can be directly compiled into a corpus. Of course, switching from some well-known word processing software to a new web interface is not automatically welcome by students and tutors alike, so they need to be offered proper "compensation" for the inconvenience. This compensation typically includes additional functions that will make the L2 acquisition easier or more entertaining. This is discussed in Sect. 4.

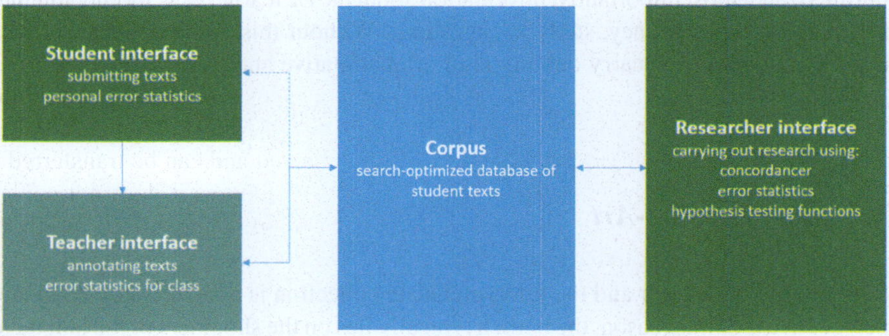

Fig. 1 A schematic drawing of the data pipeline in Hypal and Hypal4MUST

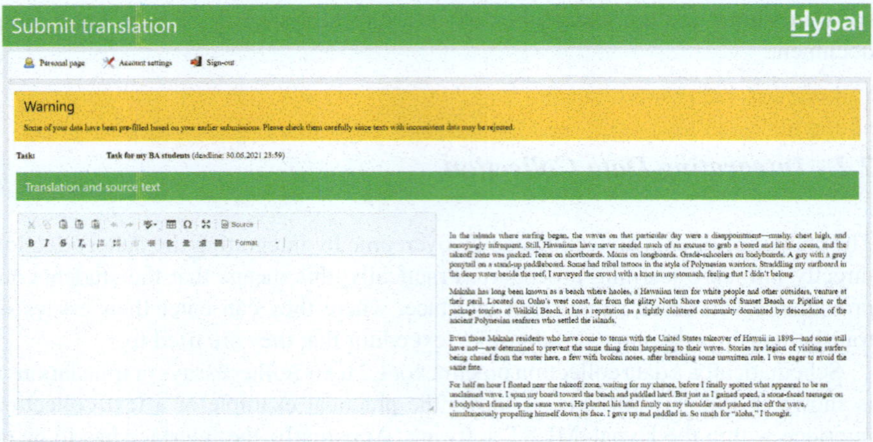

Fig. 2 A partial screenshot from the student web interface of Hypal4MUST, which shows the source text to be translated (on the right) and the rich text input window on the left, where the student is expected to enter the translation

3 Metadata Acquisition and Annotation

Apart from collecting the data in the proper markup format, LC and LTC have to be accompanied by accurate, consistent and sufficiently complex metadata. Depending on the research objectives, the metadata must cover one or more of the following classes—information about the assignment (in the case of LC) or the SL text (in the case of LTC), information about the task (e.g. in-class or homework, timed or untimed) and, most importantly, information about the L2 learner (e.g. mother tongue, self-perceived proficiency, study background). Without this kind of metadata, it is close to impossible to carry out any kind of quantitative analysis on top of LC and LTC.

3.1 State-Of-The-Art

For both the L2 learner and the tutor, metadata collection is not necessary from any perspective. For this reason, the burden typically lies on the shoulders of the language researcher, who is provided with a volume of L2 learner texts and needs to collect the necessary metadata *ex-post*. Naturally, this is a difficult and time-consuming process and in many cases, the necessary metadata cannot be even acquired.

3.2 Integrated Approach

In all fairness, being asked to supply metadata offers little advantage to the L2 learner and the tutor and it can even be viewed as an annoyance by many. For this reason, the task should ideally be distributed between the L2 learner, tutor and language researcher personas. This was achieved within the MUST project in the following way: The L2 learner needs to supply their own metadata, which typically include their language background, study background or self-perceived proficiency. It is imperative that they do not have to fill in the data every time when submitting an assignment, otherwise they might tend to go through the form as quickly as possible, entering random or distorted metadata. For this reason, the L2 learners enter most of their personal data only when registering for the platform and later on, they are only asked to update the data which change often (e.g. age, years of experience with TL). The L2 tutors are required to enter the metadata associated with the translation task. This metadata have to be entered every time when a translation task is entered and typically, they are related to the type of activity (in-class, exam or homework), timing or the possibility of using language resources. In the case of LTC, the metadata associated with the source text also have to be entered—the persona that enters the metadata is either the L2 tutor or the language researcher and the metadata includes items such as genre, reference for the ST or information about the author (Fig. 3).

3.3 Error Annotation

Error annotation is a task, in which the tutor identifies erroneous words or expressions and assigns a specific error tag to them. In some cases, the tutors may also decide to type a correction or a comment as a feedback to the student. The technical aspect is important here, as the annotation interface has to be user-friendly enough, ideally comparable to reviewing and commenting functions in text processors. However, the real challenge of error annotation is not related to the technical aspect but rather, it is related to ensuring the consistency of error annotation between various tutors and language levels. For example, some tutors may choose not to annotate some of the errors for fresh learners of the language. This fact then poses a major challenge in data interpretation but at the moment, it is outside the scope of this paper (Fig. 4).

4 The Compensation for L2 Learners and Tutors

As previously mentioned, integrating LC and LTC data collection into e-learning pipelines comes with extra effort for both the L2 learners and the tutors. For this reason, the software should offer them some kind of compensation, which will motivate them to use it.

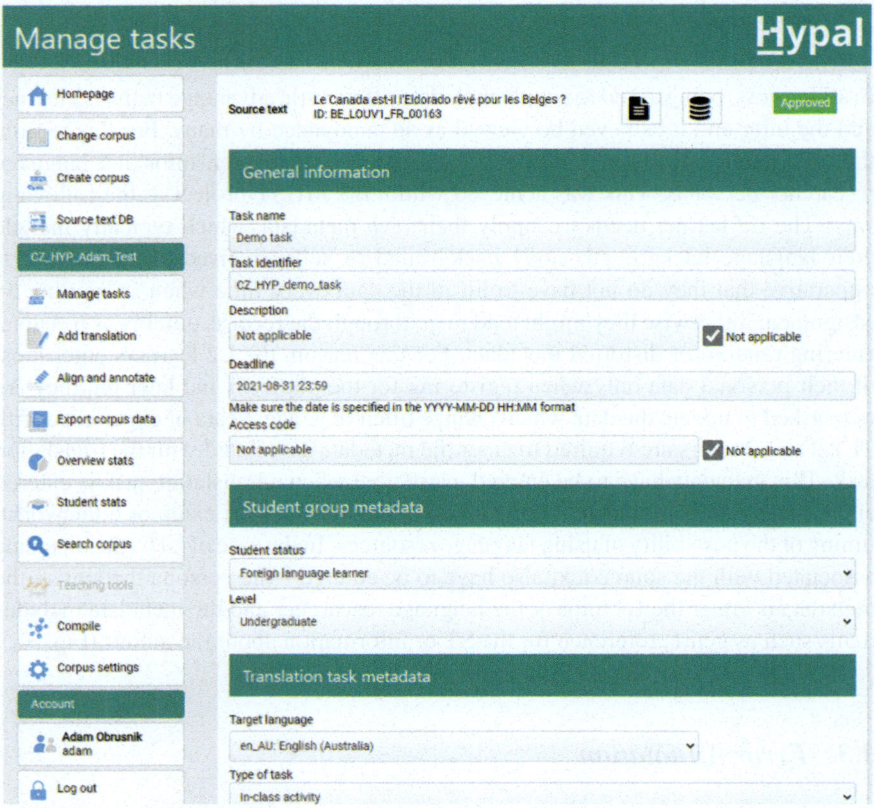

Fig. 3 Example of metadata entry form when creating a new translation task in Hypal4MUST

For the L2 tutor, it can be rewarding to be able to track the performance of their students over time. For example, with sufficient metadata, the tutor can see whether there is a measurable improvement between first year and second year students, as illustrated in Fig. 5.

For the L2 learners, on the other hand, it can be motivating to monitor their own progress during the studies. This can be done, for example, by comparing the average number of error annotations per assignment with the learner's performance in the very same translation tasks. This is illustrated in Fig. 6. Alternatively, the aggregated error annotation data can provide the learners insight into their most common error types, thereby advising them what types of errors they should focus on in their further learning. An example of how such data can be presented to the student is shown in Fig. 7.

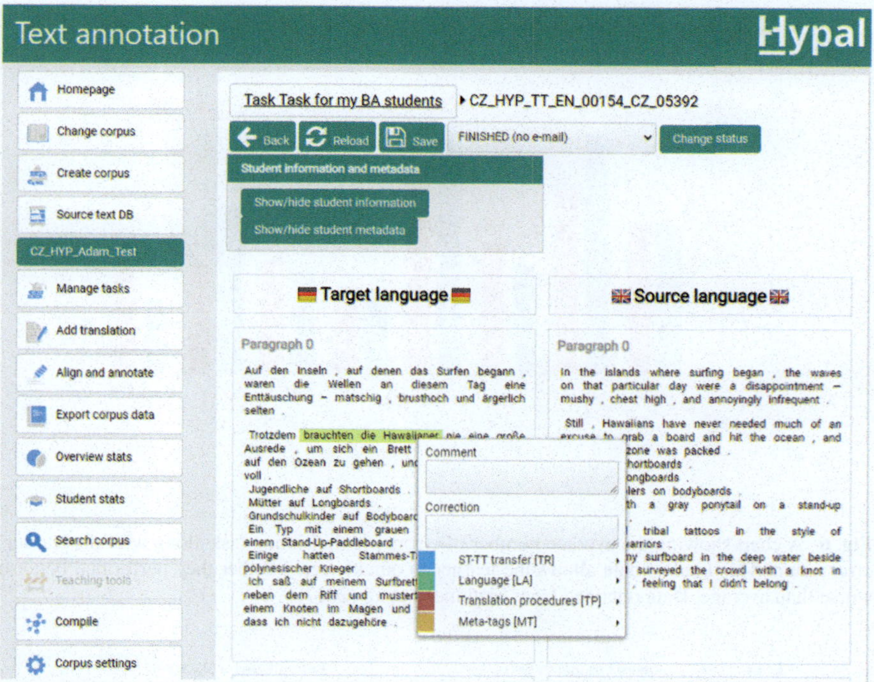

Fig. 4 A screenshot of a user-friendly annotation interface, available to the L2 tutor for highlighting the errors in learner texts or translations

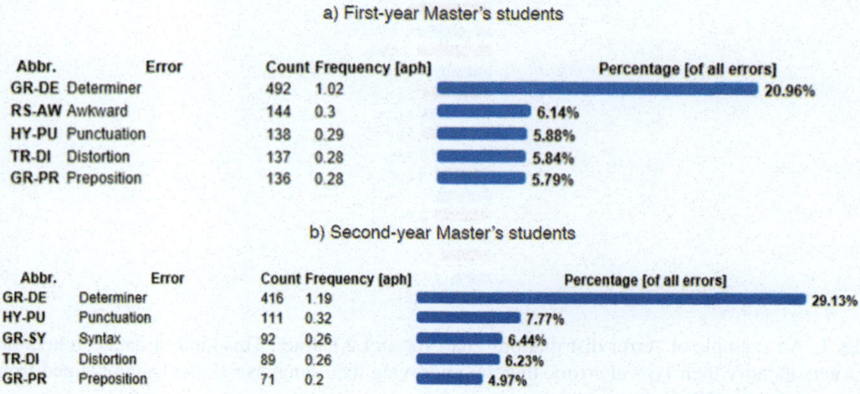

Fig. 5 Example of most common error types for two sub-groups of L2 learners. Data obtained from Fictumová et al. (2017)

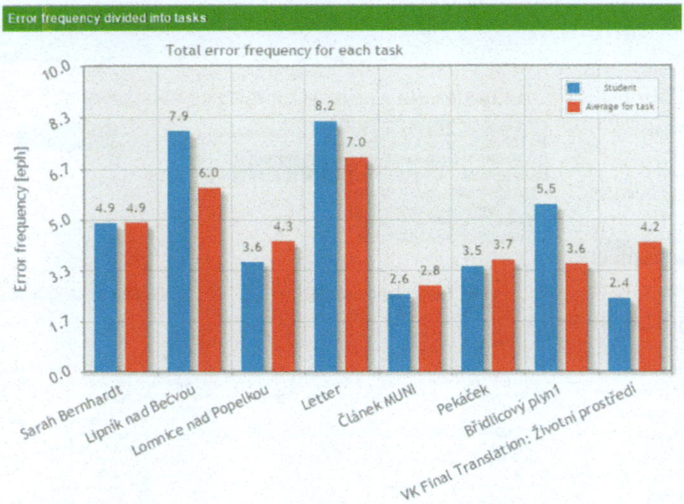

Fig. 6 A chart showing the average number of errors per translation task along with the learner's error count. This kind of data allows the learner to quickly see whether they performed better or worse than average. Data obtained from Fictumová et al. (2017)

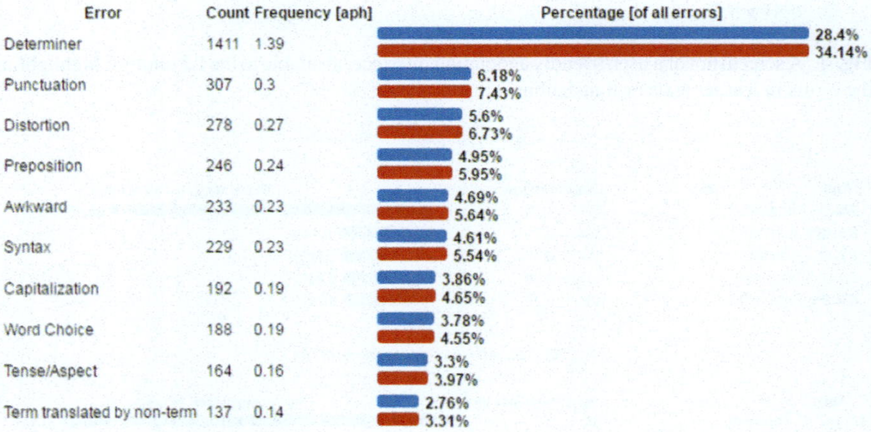

Fig. 7 An example of "error distribution" chart for an L2 learner. This kind of data can help the learners identify their typical errors, thereby improving their language skills. Data obtained from Fictumová et al. (2017)

5 Conclusions

This publication outlined the main technical challenges related to the acquisition of high-quality data for learner corpora and translation learner corpora. It aimed to illustrate that well-designed and user-friendly technical solutions can greatly facilitate the data acquisition, thereby leading to higher-quality LC and LTC. The author is currently developing Hypal and Hypal4MUST software which is capable of integrating data collection, data processing and data exploitation into a single interface and at the moment, he is starting to work on a new version of the software, which will be available under an open source license.

References

Fictumová, Jarmila, Adam Obrusnik, and Krystina Štěpánková. 2017. Teaching specialised translation. Error-tagged translation learner Corpora. *Sendebar* 28: 209–241.

Granger, Sylviane. 2017. Learner Corpora in foreign language education. In *Language, Education and Technology*, 427–440. Springer International Publishing. https://doi.org/10.1007/978-3-319-02237-6_33

Granger, Sylviane, and Marie-Aude. Lefer. 2020. The multilingual student translation corpus: A resource for translation teaching and research. *Language Resources and Evaluation* 54 (4): 1183–1199. https://doi.org/10.1007/s10579-020-09485-6.

Laviosa, Sara. 2012. The corpus-based approach: A new paradigm in translation studies. *Journal Des Traducteurs* 43 (4): 474. https://doi.org/10.7202/003424ar.

Adam Obrusnik has a B.A. in linguistics and a Ph.D. in computational physics. Although his primary career is in physics, he has been consistently working on the development of software tools enabling user-friendly data acquisition and pre-processing for learner corpora and translation learner corpora. The software tools called Hypal and Hypal4MUST have been subject to active development for the past 7 years. At the moment, Adam is exploring the possibility of developing a more modern and modular version of the software.

Investigating the Chinese and English Language Proficiency of Tertiary Students in Hong Kong: Insights from a Student Translation Corpus

Jun Pan, Billy Tak Ming Wong, and Honghua Wang

1 Introduction

The cultivation of bilingual (i.e. Chinese and English) personnel has long been a primary goal of education in Hong Kong. Despite the significance of bilingual proficiency enhancement, much remains unknown as to what aspects of tertiary students' bilingual proficiency should be enhanced.

Translation, as one of the prerequisites of bilingual competence, is often used to test and demonstrate the level of one's language proficiency. It has also been used in foreign language teaching as one of the earliest pedagogical tools. Therefore, students' translations consist of valuable data for the study of bilingual proficiency.

This study aims to investigate the Chinese and English language proficiency of tertiary students in Hong Kong through the unique lens of translation. An error-annotated translation learner corpus—the Hong Kong subset of the Multilingual Student Translation (MUST) corpus—was developed following the standard of an international multilingual corpus initiative for the study of translated language of language learners and translation students worldwide (Granger and Lefer 2017).

Tapping into the standardised error annotation scheme and rich contextual information of the source texts, student translators, translation tasks, etc. of the MUST

An earlier draft of this chapter was submitted to the funding body as part of the final project report.

J. Pan (✉)
Hong Kong Baptist University, Kowloon Tong, Hong Kong, China
e-mail: janicepan@hkbu.edu.hk

B. T. M. Wong
Hong Kong Metropolitan University, Ho Man Tin, Hong Kong, China
e-mail: tamiwong@hkmu.edu.hk

H. Wang
The Hang Seng University of Hong Kong, Siu Lek Yuen, Hong Kong, China
e-mail: ansonwang@hsu.edu.hk

J. Pan and S. Laviosa (eds.), *Corpora and Translation Education*, New Frontiers in Translation Studies, https://doi.org/10.1007/978-981-99-6589-2_9

initiative (ibid), the study built an error-annotated learner corpus of over 300,000 word tokens that helps to unveil the problematic aspects in the Chinese and English language proficiency of tertiary students in Hong Kong and pinpoint the most urgent problems for improvement. The study also sheds light on the design of language proficiency enhancement strategies catering for the needs of students at tertiary institutions.

2 Research Background

2.1 *The Language Education Policy and Bilingual Proficiency of Students in Hong Kong*

Hong Kong nurtures a unique language environment, where both Chinese and English have been stipulated as its official languages since the United Kingdom transferred its sovereignty back to the People's Republic of China (PRC) in 1997.[1] The 1997 turnover led to a lot of discussion and debate on Hong Kong's language education policy (Evans 2013; Lin and Man 2009), which were thereafter ratified as "biliterate and trilingual".[2] The Education Bureau of Hong Kong has spelled out the constitution and current interpretation of this language policy as follows:

> The language education policy of the Government of the HKSAR aims to enable our students to become biliterate and trilingual. We expect that our secondary school graduates will be proficient in writing Chinese and English and able to communicate confidently in Cantonese, English and Putonghua.[3]

To achieve the goal of cultivating biliterate/bilingual[4] talents, many studies were carried out on the implementation and effectiveness of schools' policies of medium of instruction (MOI) (see Evans 2013; Lin and Man 2009). Most studies indicated the difficulties in employing English medium of instruction (EMI) in local secondary schools, largely attributable to the inadequacy of students' English language proficiency and lack of school/teacher support (Lin and Man 2009). However, the Chinese medium of instruction (CMI) policy, prevalently employed after 1997, was not well received, mostly due to its mismatch with the EMI policy adopted widely in local higher education institutions and the lower prestige of the mother tongue in Hong Kong (Evans 2013; Lin and Man 2009). Lin (2015) therefore proposed the systematic use of L1 (Chinese) in bilingual classes focusing on content and language integrated learning (CLIL).

---

[1] See GovHK website: https://www.gov.hk/en/about/abouthk/facts.htm.

[2] See GovHK website: http://www.policyaddress.gov.hk/pa99/english/espeech.pdf.

[3] See Education Bureau's website: http://www.edb.gov.hk/en/edu-system/primary-secondary/applicable-to-primary-secondary/sbss/language-learning-support/featurearticle.html.

[4] *Bilingual* in this study refers to *biliterate*, since this study concerns only the written Chinese and English. Similar uses can also been found in Lin and Man (2009).

In addition to the general discussions and theoretical considerations, a number of empirical studies were conducted to investigate the relationship between MOI and bilingual or monolingual development of students at primary or mostly secondary schools. Tsang (2008), for instance, found that junior-form students from CMI schools, although obtained higher scores in integrated content subject learning, had lower achievements in English language learning than junior-form EMI students. In addition, senior-form students, whilst stopping to benefit from the positive effect of CMI from their junior-form education on content subject learning, continued to get lower scores in English language, and had smaller chances to enter tertiary level education (which often employs EMI) than their EMI peers (also see Evans 2013; Lin and Man 2009). Another study, Lo and Lo (2014), through a meta-study of 24 empirical studies, indicated that the application of EMI in secondary schools, whilst successfully contributing to higher levels of students' English language proficiency, also led to lower levels of Chinese language proficiency and insufficient command of the content knowledge.

Nevertheless, few studies provided insights into specific aspects for the improvement of students' written Chinese and English proficiency (i.e. the "biliterate" goal). In this regard, Lin and Morrison (2010) tested the impact of MOI in secondary schools on tertiary students' English academic vocabulary, which constituted a key player of students' academic achievement at the tertiary level. Through comparing their results with Fan's (2001) study carried out at the beginning of the MOI policy change, Lin and Morrison identified a significant decrease in the size of students' English academic vocabulary, partially assignable to the increase of CMI secondary schools. These findings are useful not only for the review of current CMI policy for secondary schools, but also because they help to pinpoint the special aspects that are worthy of attention.

With the Government's call for "fine-tuning" its language policy (Education Bureau 2010),[5] there is a need for a systematic investigation of the problematic aspects in current students' bilingual proficiency, upon and after the end of their secondary study. However, such an endeavour is yet to be addressed in the literature.

2.2 *Translation and Language Education*

The relationship between translation and language education is long-standing and far-reaching. Translation was, in the first place, employed as one of the earliest methods of foreign/second[6] language (FL/L2) teaching, i.e. the Grammar-Translation Method, at the outset in the teaching of classic languages of Greek and Latin back in the sixteenth century (Richards and Rodgers 2001). With the development of translation studies

[5] See Education Bureau's website: http://www.edb.gov.hk/attachment/en/edu-system/primary-secondary/applicable-to-secondary/moi/2nd_moi_booklet.pdf.

[6] The terms *foreign language (FL)* and *second language (L2)* teaching are used interchangeably in the study.

as an academic discipline, scholars have been calling for differentiating between *translation in language teaching* and *language teaching for translators*—the former treats translation as a "significant component" in language teaching, and the latter focuses on "how translation might most effectively be provided with the kinds of linguistic skills which will help foreign language learners produce socio-functionally adequate texts in the most economic quality-oriented manner possible" (Malmkjær 1998, pp. 1–2).[7] Meanwhile, the use of translation encountered "rejection" in L2 teaching since the introduction of the Direct Method for language teaching, i.e. the use of only the target language in foreign language classrooms, was introduced towards the turn of the twentieth century (Cook 2010).

Although the use of translation in language teaching has experienced ebbs and flows in history, there has been a revived interest in the indispensable relationship between the two, especially in higher education (Laviosa 2014):

> Since the turn of the century, the debate about the merits of translation as a method of language learning, teaching and testing has been enriched by critical reflections on the value of educational translation as an aid to second language acquisition, as a means of developing metalinguistic competence, as a motivational factor, as an essential skill in today's multilingual societies and globalised world and as an ecological practice that not only recognises the value and relevance of students' first language but also facilitates the creation of multilingual identities and protects linguistic as well as cultural diversity. (p. 28)

This revival of recognising the role that translation plays in language teaching corresponds well with the recent revitalisation of bilingual or mixed MOI (see Lin 2015), as discussed in the previous section.

Whilst many studies naturally press for the use of translation as a means for L2 teaching or enhancement (see Laviosa 2014), the application and benefits of translation in first language (L1) education have also been, although far less frequently, touched upon. Horner and Lu (2012) suggested that translation, as a translingual approach, can be employed in tertiary-level English writing classes in the United States, whereby both native and non-native English speakers can collaboratively improve their understanding of writing in a wider sense. They extended the notion of "teaching writing in English" to "rewriting English". Their translingual approach features the construction of multilingual identity and preservation of cultural diversity mentioned in Laviosa (2014).

In addition, Ngan (2009) addressed the relevance between bilingualism and translation, and proposed the incorporation of the bilingual representation method in biliteracy training. The author defines bilingual representation as "a complicated process which involves selecting from the TL[8] corresponding counterparts of the SL with reference to the use of the SL in the context of the source text (ST)" (p. 41), a method often employed in practical translation. Moreover, Sidiropoulou (2015), focusing

---

[7] This study focuses on *language learners* instead of *translation learners*. Therefore, it relates more to *translation for language learning* rather than *language teaching for translators*. However, these two are not entirely indispensable as suggested in the revived yet mutually enriching relationship of the two disciplines of translation and language teaching (see Cook 2010; Laviosa 2014).

[8] SL refers to the source language in translation, and TL the target language. Likewise, ST stands for source text and TT for target text.

on modal markers, illustrated the usefulness of translation-related parallel data in foreign language teaching.

Apart from the pedagogical application of translation in language teaching, the notions of linguistic competence and translation competence are mutually inclusive. Linguistic competence is naturally included in the components of translation competence (PACTE 2003, p. 58) as "the underlying system of knowledge needed to translate".

In a different manner, language learning/teaching, in particular L2 teaching, implies a translational component. The notion of "competence" in language learning/teaching has been extended from Chomsky's (1965, p. 4) "linguistic competence" to "communicative competence" (Hymes 1972). The former referred to "the speaker-hearer's knowledge of his language" (cf. the notion of "performance", i.e. "the actual use of language in concrete situations", Chomsky 1965, p. 4). The latter was further divided into linguistic competence, sociolinguistic competence, discourse competence, and strategic competence (Canale and Swain 1980). Communicative language teaching (CLT) was therefore built upon these competence components, within which, translation (also interpreting) was taken as "the fifth skill", in addition to the four basic skills of reading, writing, listening, and speaking in L2 teaching (Naimushin 2002). Moreover, in Selinker's (1992) model of interlanguage (IL, i.e. the language produced by L2 learners) competence, translation skill is taken as an important indicator of L2 competence.

Furthermore, translation has been used widely in language assessment (Tsagari and Floros 2013). Ricardo-Osorio (2008), through a survey of FL learning outcomes assessment methods of undergraduate programmes in the United States, showed that translation was the fourth most widely used assessment method, following faculty designed tests, student papers, and student presentations. Likewise, Sun and Cheng (2013), through an empirical studies, found that translation was a valid measure for students' FL competence.

2.3 Learner Corpora and Language Learning

The study of learner language has long been an important aspect in language learning research, including both L1 and L2 acquisition as well as bilingual development (see Poulisse 1999). At the core of learner language study is what Corder (1967/1983) refers to as "the systematic errors of the learner from which we are able to reconstruct his[/her] knowledge of the language to date, i.e. his[/her] *transitional competence*" (p. 168). This notion of language errors, due to its limitation of only addressing the static and negative side of learner language, was later developed into the IL hypothesis, which addresses learner language from a developmental point of view, defined by Selinker (1972/1983) as "a separate linguistic system based on the observable output which results from a learner's attempted production of a TL [Target Language, in this case the second language the learner is attempting to learn] norm" (p. 176, elaborations added).

As discussed in the previous section, translation is an important indicator of IL competence (Selinker 1992). Al Khafaji (2007) compared translation to translanguage, i.e. "a transitionally unstable linguistic entity that evolves during acts of translation along intersecting stages in a 'trip' stretching from the ST towards the TT during which hybrid 'language' comes into being banking on the linguistic and social potentials of the SL and TL" (p. 473). Therefore, translation, in the sense of translanguage, is reflective of the (in)competence in a SL and TL.

The development of corpora has contributed greatly to the research and practice of language learning. Steward et al. (2004) identified three primary aspects linking language corpora and language learners. The primary one is *corpora by learners*, defined as the development of corpora that "can be used to study features of interlanguage" (p. 2). The second aspect is *corpora for learners*, referring to those "designed to benefit learning by allowing teachers and material designers to provide better descriptions of the language to be acquired" (p. 6). The last concerns *corpora with learners*, relating to "activities designed to help learners use corpora and to acquire linguistic knowledge and skills through their use" (p. 8).

The study of corpora by learners has great potential for the investigation of "the systematic errors of the learner" (Corder 1967/1983, p. 168), or the systematic analyses of the IL, and can therefore shed light on the development of focused "teaching methods and contents … so as to speed acquisition" (Steward et al. 2004, p. 3). Granger (2002), in particular, gives a definition of "learner corpora" in FL/L2 learning that can be extended to L1 learner corpora as well:

> Computer learner corpora are electronic collections of language textual data assembled according to explicit design criteria for a particular language teaching purpose. They are encoded in a standardised and homogeneous way and documented as their origin and provenance. (Adapted from Granger 2002, p. 7)

Nevertheless, the greatest challenge in learner corpus research lies in "identifying and classifying errors, and hypothesising 'correct' version corresponding to the learner's intentions" (Steward et al. 2004, p. 3). The lack of a consistent and comprehensive error classification system and the painstaking efforts involved in the annotation process may have led to the limited progress in learner corpus research (also see Granger 1998).

Despite this difficulty, pioneering efforts have been made for learner corpus research. One of the most significant outcomes is the International Corpus of Learner English (ICLE; see Granger 1998; Granger et al. 2009), the "best-known" learner corpus (McEnery et al. 2006, p. 66) with 3.7 million English words in size composing of essays written by advanced English learners from 16 different L1 backgrounds (Granger et al. 2009).[9] ICLE features a standardised learner profile questionnaire and an error annotation scheme designed specifically for the language learners (ibid). The corpus has helped to greatly advance the study of learner language and the development of learner-corpora-informed L2 teaching resources (Granger 2003; Steward et al. 2004), for example, the Teachers of English Education Nexus (TeleNex), as

[9] See the ICLE website: http://www.fltr.ucl.ac.be/fltr/germ/etan/cecl/Cecl-Studys/Icle/icle.htm.

"a computer network providing continuous professional support to English language teachers in Hong Kong primary and secondary schools".[10] The study website includes both data of student problems and teaching implications (see Granger 2003).

Apart from the ICLE, other major international initiatives of learner corpus include the Longman Learners' Corpus, with 10 million English words written by English learners from 20 different L1 backgrounds, well-known for its use in dictionary and course book compilation that addresses "students' specific needs",[11] and the Cambridge Learner Corpus (CLC), a learner corpus collecting written English from 250,000 language learners all over the world, including those produced by those taking the Cambridge ESOL English exams.[12] In addition, there are also a few L2 learner corpora with data collected only from Chinese speakers, including the Taiwanese Learner Corpus of English (Shih 2000), the Chinese Learner English Corpus (CLEC; Gui and Yang 2003), the Spoken and Written English Corpus of Chinese Learners (SWECCL; Wen et al. 2005), and the College Learners' Spoken English Corpus (COLSEC; Yang and Wei 2005). These aforementioned L2 learner corpora usually feature learner language collected from post-secondary students.

In a different manner, L1 learner corpora mainly focus on the aspect of children language development (Behrens 2008). Examples of major corpora in this regard include the Child Language Data Exchange System (CHILDES),[13] the Polytechnic of Wales (POW) Corpus,[14] and the Lancaster Corpus of Children's Study Writing (LCPW).[15] Amongst them, LCPW, which contains longitudinal data taken from 37 children aged between 9 and 11 in the United Kingdom, is the only one that focuses on written language. In addition, CHILDES also features a sub-corpus of data provided by bilingual children.

To conclude, although a learner corpus can provide valuable insights into learner language features that can be utilised in the development of teaching resources (e.g. teaching materials, dictionaries, and online learning platforms) tailored to specific learner needs, its development is still limited largely by the difficulties inherent in standardised data collection method and annotation schemes. Most of the existing learner corpora focused on L2 language learners, although a few of them were on children's L1. Whilst the L1 and L2 learner corpora were not comparable due to inconsistent data annotation schemes and different student levels, the bilingual children data in CHILDES involve only spoken language (of 1 child in Hong Kong) and

---

[10] See the TeleNex website: http://www.telenex.hku.hk/telec/pmain/opening.htm.

[11] See the Longman Learners' Corpus website: http://www.pearsonlongman.com/dictionaries/corpus/learners.html.

[12] See the CLC website: http://www.cambridge.org/gb/cambridgeenglish/better-learning/deeper-insights/linguistics-pedagogy/cambridge-english-corpus.

[13] See the CHILDES website: http://childes.talkbank.org/.

[14] See the POW website: http://clu.uni.no/icame/manuals/POW.HTM.

[15] See the LCPW website: http://www.lancaster.ac.uk/fass/studys/lever/.

has limited scope of application and implication. There seems to be no readily applicable corpus that can be used to study the learner language features in the biliterate setting in Hong Kong.[16]

Thus in this study, an error annotated learner corpus, i.e. the Hong Kong Student Translation (HKST) corpus, was developed with the purpose to sample learner language from tertiary institutions in Hong Kong. The corpus was developed as part of and based on standards developed for an international initiative of multilingual student translation (MUST) corpus (Granger and Lefer 2017, 2020). The MUST project adapts the framework for developing ICLE, and aims to build a large multilingual student translation corpus with the collaborative efforts of researchers from different parts of the world (Granger and Lefer 2017, 2020). The collaborative initiative aims to cover, at the time of writing this chapter, 25 languages and 50 language pairs, wherein the first author of this chapter serves as a regional coordinator.[17] Situated at the intersection of learner corpus research and corpus-based translation studies, the MUST initiative features the collection of rich contextual information about the learners' backgrounds and translation settings, as well as a shared annotation scheme of language errors for both learner language and translation research (ibid).

Apart from the MUST framework, the study also taps into relevant developments in corpora and translation education (Pan 2019a, 2021a, b), as well as previous work by members of the corpus compilation team on Chinese/English language learning (Yan and Pan 2016), the relationship between learner variables and learner performance, including learning achievements and problems, in language learning and translation/interpreting training (Pan 2012, 2014; Pan and Wang 2012; Pan and Yan 2012, 2014; Yan et al. 2010; Yan and Wang 2012, 2015), corpus compilation (Chow and Wong 2015; Pan and Wong 2017; Pan et al. 2022), in particular, translation/interpreting learner corpus design (Pan 2012, Pan and Chan 2013; Pan 2017; Yan and Wang 2014; Pan et al. 2022), the application of linguistic features for text quality assessment (Wong 2010; Pan et al. 2022), linguistic annotation of corpora data (Pan and Wong 2015a, b, 2017; Wong and Lee 2013; Pan et al. 2022), computer tool development for semi-automatic annotation of linguistic features (Wong and Lee 2013; Wong et al. 2014; Chow and Wong 2015), and language and identity (Chan and Fong 2016). The development and periodic reports of the study findings have helped to testify the feasibility of a large-scale corpus for the study of language proficiency of tertiary students in Hong Kong (Pan and Wang 2017, 2018; Pan 2019b; Pan and Wong 2021; Pan et al. 2021a, b, 2022).

---

[16] It should be mentioned that the Chinese/English Translation and Interpreting Learner Corpus (CETILC, Pan et al. 2022), developed by the authors of this chapter for a different project, can complement the corpus developed in this study, with a special focus on the spoken and written outputs produced by translation major students in Hong Kong.

[17] More information can be obtained from the MUST website: https://uclouvain.be/en/research-institutes/ilc/cecl/must-partners.html.

3 The Study

This study aims to investigate the Chinese and English language proficiency of tertiary students in Hong Kong through the unique lenses of translation. Corpus compilation and annotation constituted two major steps in the investigation. Employing instruments of contextual/learner data collection and error annotation scheme developed for the MUST international initiative (Granger and Lefer 2017, 2020), this study examines the carefully collected learner translation outputs by tertiary students in Hong Kong using both quantitative and qualitative methods.

In particular, the study pivots on two main research questions:

1. What are the high-frequency error types in written Chinese/English of tertiary students in Hong Kong?
2. What are the relationships between the types of Chinese/English language features and relevant contextual/learner factors?

3.1 Corpus Compilation

The HKST corpus, i.e. the Hong Kong subset of the MUST corpus, consists of translations and metadata provided by students from more than 11 tertiary institutions in Hong Kong. Six main batches of data collection were performed during the study period (Sep 2018–Aug 2021).

Figure 1 displays the self-reported data of students participating in the latest batch of data collection.

The participating students included undergraduate Chinese and English language learners enrolling in courses at language centres, and Chinese and English language departments (including but not limited to translation students). Language centres usually aim at helping students with their effective writing in both the Chinese and English languages.[18] They usually provide institution-wide credit- and non-credit-bearing courses to students from different departments of the institution.[19] In some institutions, Chinese/English language departments also undertake the work of language centres in providing institution-wide language courses.[20] In this study, students from both types of institution-wide language courses were invited to contribute data, and constituted the primary participant group, coded as Chinese/English language general learners (CLGLs/ELGLs).

Chinese and English language departments (including translation programmes/departments) normally have specific criteria for the recruitment of students on the basis of their Chinese and English grades in the Hong Kong Diploma of Secondary

[18] See, for example, the HKBU language centre website: http://lc.hkbu.edu.hk/mission.php.

[19] See, for example, the HKBU language centre website: https://lc.hkbu.edu.hk/main/chinese-course/.

[20] See, for example, the HSMC Chinese department website: https://chi.hsu.edu.hk/programme/common_core/.

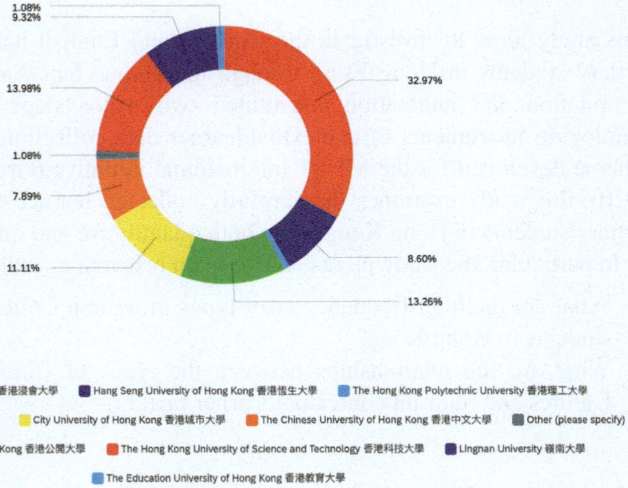

Fig. 1 Student participants of the study (latest batch)

Education (HKDSE). Therefore, student participants from these departments made up the second group in the study, coded as Chinese/English language major learners (CLMLs/ELMLs).

Figure 2 shows the percentage of the student majors.

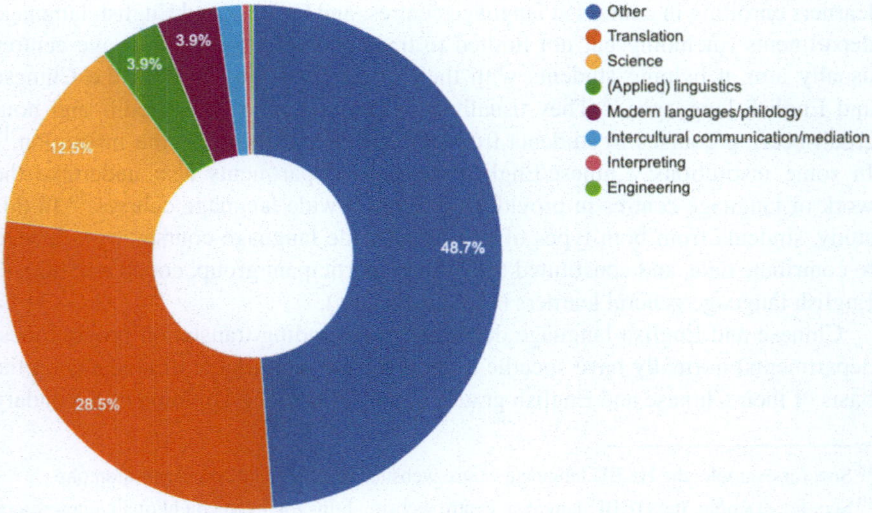

Fig. 2 Current study backgrounds of the participating students

The participating students were asked to provide translations in both directions, from Chinese to English and from English to Chinese. The texts for translation were on general topics, including selected excerpts taken from newspaper or magazine articles (Table 1). Each of the text was about 250–600 words in length, in line with the MUST specifications (Granger and Lefer 2017). The investigators of the study assessed the text in terms of translation difficulty level based on their expertise of translation whilst taking into consideration basic criteria such as type/token ratio and the vocabulary range. The array of texts has the potential to enlist a wide range of language errors.

Contextual information of the texts were coded by the Principal Investigator, also the first author. Figure 3 shows the codes used for a sample source text.

Student translations were collected through the tailor-made Hybrid Parallel Text Aligner for the MUST corpus, i.e. Hypal4MUST (Granger and Lefer 2017; Fig. 4). Each translation took about 40–60 min.

Apart from the translation data, metadata information of each corpus entry (Fig. 5), including student and task specific data (Granger and Lefer 2017; Pan and Wang 2017) were also collected through a survey and uploaded to the same platform. Students took about 15 minutes on average to complete the information.

The collected data were further processed for data cleansing and parallel text alignment on the Hypal4MUST platform, with the goal of pairing up the Chinese–English bilingual texts at the sentence level.

Table 1 The list of source texts used for the corpus

Tears, fears and cheers: How did your workplace handle the post-election fallout?
Moonlight's Barry Jenkins on Oscar fiasco: "It's messy, but kind of gorgeous"
Trump delays a tariff deadline, citing progress in china trade talks
Green Book' Review: A Road Trip Through a Land of Racial Clichés
賈寶玉的大紅斗篷與林黛玉的染淚手帕《紅樓夢》後四十回的悲劇力量 (Jia Baoyu's red cloak and Lin Daiyu's tear-stained handkerchief: the tragic power of the last forty chapters of *Dream of the Red Chamber*)
Lowering bar for disadvantaged students has failed to redress imbalance in university admissions, regulator says
A company's meeting on its volunteering projects
好好過日子——時間沒有溜走 (Live a good life—time hasn't slipped away)
内地「碼農」的覺醒——抗議「996」還我加班費 (The awakening of "code farmers" in the Chinese mainland—protesting "996", give me back my overtime pay)
The Guardian view on extinction: time to rebel
好好過日子——藥不能停 (Live a good life—the medication cannot stop)
Overcome procrastination
心寬, 路更寬 (With a broad mind, the road becomes wider)

```
source_text_author_speaker_status="Native speaker of the SL"
source_text_publication_status="Published" source_text_title_full_text="Tears, fears & cheers:
How did your workplace handle the post-election fallout?" source_text_author_name="Pat
DiDomenico " source_text_sampling="Sample" source_text_communicative_purpose="Mixed
communicative purpose" source_text_keyword2="workplace"
source_text_publication_format_subtype="Web"
source_text_reference_translation_f="file_I5IEYMM2SLBB.pdf"
source_text_reference_translation_publication="Unpublished at the time of data collection"
source_text_whole_text_f="file_B21525IKW0XD.pdf"
source_text_reference_translation_directionality="L2 to L1" source_text_author_type="Sole
author" source_text_language_type="General language" source_text_title_segment="Tears, fears &
cheers: How did your workplace handle the post-election fallout?"
source_text_target_audience="Specialized external audience"
source_text_general_language_type="Journalistic texts"
source_text_opinion_article_subgenre="Commentary/personal column"
source_text_whole_text_available="Yes" source_text_reference_translator_status="Member of
teaching/research staff" source_text_st_translation_status="Original text"
source_text_keyword3="emotions" source_text_publication_date="2016-11-18 23:59"
source_text_keyword1="presidential election" source_text_reference_translation="Available"
source_text_language="en" source_text_publication_url="https://www.businessmanagementdaily.com/
47609/tears-fears-cheers-how-did-your-workplace-handle-the-post-election-fallout"
source_text_publication_format="Electronic document/digital copy"
source_text_publication_reference="Not applicable" source_text_sample_position="Beginning
sample" source_text_mode="Written to be read" source_text_length_words="252"
source_text_journalistic_texts_genre="Opinion article"
source_text_reference_translation_target_language="zh_Hant_HK" >
```


Fig. 3 Metadata coded for a sample source text

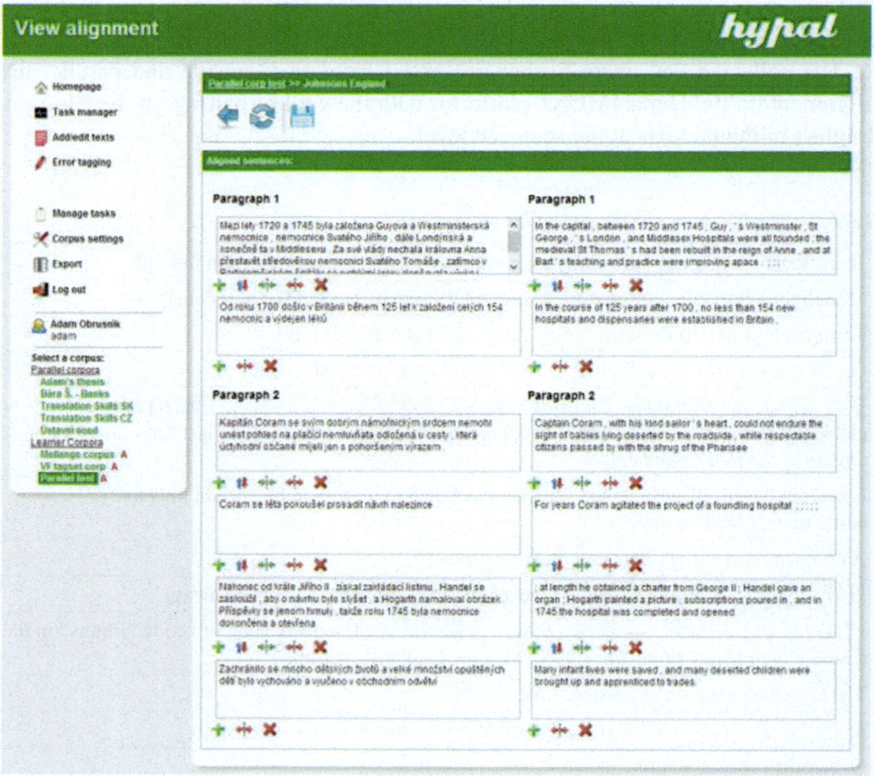

Fig. 4 The Hypal interface (Obrusník 2014, p. 68)

```
student_mother_tongue="zh_Hant_HK" task_student_tl_status="Mother tongue"
student_experience_with_cat="No" student_gender="Male" student_main_language_university_1="en"
student_main_language_university_2="na" student_main_language_university_3="na"
student_foreign_language_2="zh_Hans" student_foreign_language_3="es"
student_main_language_secondary_school_1="en" student_foreign_language_1="en"
student_sl_proficiency="Intermediate" task_student_duration="60" student_age="20"
student_translation_internships="No" student_semesters_studying_translation="2"
task_student_feedback_type_before_submission="Individual teacher's feedback"
task_student_tools_resources="Bilingual dictionaries"
task_student_feedback_before_submission="Yes" student_current_study_background="Translation"
student_tl_proficiency="Native" student_mother_tongues="1 mother tongue"
student_sl_countries_stays="No" student_prior_study_background="Other"
student_main_language_secondary_school_2="zh_Hant" student_sl_years_studying="17"
task_student_sl_status="Foreign language" student_main_language_primary_school_1="en"
student_main_language_primary_school_2="zh_Hant_HK" translation_brief_file_available="Yes"
marking="Marked" student_status="Trainee translator" translation_brief_additional_info="Not
applicable" level="Undergraduate" type_of_task="Examination" translation_brief_used="Yes"
translation_brief_audience_specified="Not specified" translation_brief_layout_and_guide="No
instructions given" translation_brief_requester_specified="Not specified" duration_minutes="60"
translation_brief_use_glossaries="No instructions given"
translation_brief_f="translation_brief_fileN75RTTMS1YMV.pdf" translation_brief_format="Hard copy"
translation_brief_format_specified="Specified" duration="Timed"
translation_brief_communicative_purpose_specified="Not specified" revision="Only one version of
the TT is expected" translation_brief_use_memories="No instructions given"
tools_and_resources="Allowed" target_language="zh_Hant_HK"
```


Fig. 5 Metadata of a sample student and translation task information

3.2 *Corpus Annotation*

Apart from POS tagging, the corpus was annotated with errors made in the student translations according to a standardised three-layer error annotation scheme of the MUST initiative, i.e. the Translation-oriented Annotation System (TAS 1.0), which was developed by drawing on a diverse array of prominent error schemes from both language and translation studies worldwide (Granger and Lefer 2017, 2020).

The version of the MUST annotation scheme used for the study integrated several major frameworks. The CELTraC error typology (Fictumová et al. 2017), specifically developed for the annotation of translation learner corpus was incorporated as one of the main frameworks. The typology took into consideration transfer and language errors on a two-layer system and was already incorporated to the Hypal interface (Granger and Lefer 2017; Fig. 6). Its primary annotation categories included content transfer, grammar, terminology and lexis, hygiene, and register and style (ibid).

Another major framework was the Université catholique de Louvain Error Editor (UCLEE), a three-layer scheme used for the annotation of errors in FL student writing. The primary error categories included form, grammar, lexis, punctuation, sentence, word, lexico-grammar, and infelicities (Granger and Lefer 2017).

In addition, the annotation scheme took into consideration partner discussions at the series MUST workshops (2016–now). In the end, TAS 1.0 included the following categories (Table 2).

During the annotation process (Fig. 7), the study team first piloted the annotation scheme on a small sample of the Hong Kong corpus to validate their suitability for the data collected. The Principal Investigator then trained the annotators on the annotation scheme. Sample annotations and discussions were made within the study team to make sure all annotators understood the annotation scheme correctly and consistently. Then annotation was performed. When different annotators worked on the same task, comparison and training were performed to help reach an initial inter-annotator

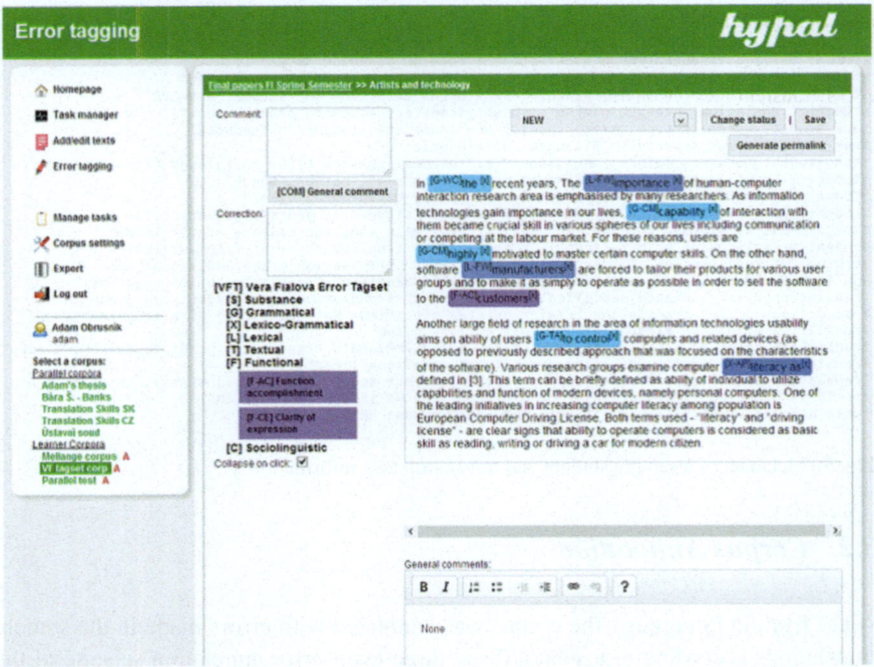

Fig. 6 The Hypal Error tagging interface (Obrusník 2014, p. 68)

reliability of up to 98%. Adjustments to the annotation scheme and annotation logs were recorded along the process (Pan and Wong 2021). The annotation was performed and recorded on the Hypal4MUST platform.

3.3 Corpus Analysis

The corpus data were then analysed through the Hypal4MUST platform and the corpus analysis software Sketch Engine (Kilgarriff et al. 2014). The high-frequency errors in the Chinese and English subsets were calculated respectively, and selected learner/contextual factors were chosen as parameters for cross-comparison amongst different subsets of the corpus.

4 Results and Discussion

The upcoming sections of this paper outline the primary findings from the corpus.

Table 2 Annotation scheme (TAS 1.0, Granger and Lefer 2017, 2020)

Layer 1	Layer 2	Layer 3
ST-TT transfer (TR)	Content transfer (CT)	Omission (OMI) Addition (ADD) Distortion (DIS) Indecision (IND)
	Lexis (LE)	Translating untranslatable (TUN) Untranslated translatable (UNT) Term translated by non-term (TNT) Non-term translated by term (NTT)
	Discourse/pragmatics (DP)	Connectors (CON) Theme-rheme (THR)
	Register and culture (RC)	Register mismatch (REG) Cultural mismatch (CUL)
	Translation brief (TB)	Inconsistency with glossary (GLO) Formatting (FOR)
Langauge (LA)	Grammar (GR)	Inflectional morphology (INF) Tense/aspect (TNS) Voice (VOI) Word order (WOR) Determiner (DET) Pronoun (PRO) Preposition (PRE) Concord (CCD) Complementation (COM) Adjective (ADJ) Noun (NOUN) Verb (VRB) Adverb (ADV)
	Lexis and terminology (LT)	Single word non-term (SWN) Derivative (DER) Cognate (COG) Single word term (SWT) Multiword non-term (MWN) Compound (COP) Collocation (COL) Idiom (IDI) Multiword term (MWT)
	Cohesion (CO)	Pronoun reference (PRF) Linkword (LIN)
	Mechanics (ME)	Punctuation (PUN) Units, dates, numbers (UDN)
	Style and situational context (ST)	Heavy (HEA) Redundant (RED) Contextual variant (COV) Degree of (in)formality (FML)

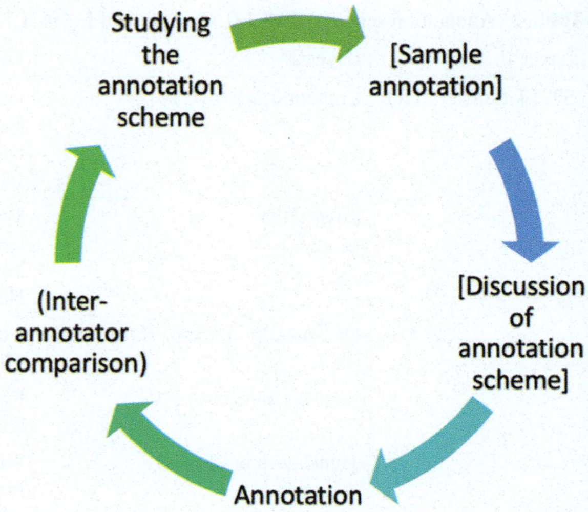

Fig. 7 The annotation process employed for the study

Table 3 Corpus statistics

	Tokens	Types	Type/token ratio (%)
Chinese	195,448	5122	2.62
English	131,295	4239	3.23
Total	326,743	9361	2.86

4.1 Corpus Statistics

Based on the calculation performed by Sketch Engine, the corpus consists of over 300,000 word tokens, with 195,448 in the Chinese subset and 131,295 in the English language subset, each with a type/token ratio of 2.62–3.23% (Table 3).

4.2 Most Frequent Error Tags in the Chinese Sub-corpus

Figure 8 shows the most frequent error tags in the Chinese sub-corpus. Distortion was, apparently, the highest frequency error type, which was mostly triggered by misunderstanding of the source language, as well as inaccurate target language expression. Nouns and verbs were the most common parts-of-speech in which distortion occurred (Fig. 9). When language errors were taken into consideration, heavy structure (style and situational context), multiword non-term collocation (lexis and terminology), and pronoun reference (cohesion) were the top three problems amongst the students. These problems are the urgent issues that both secondary and tertiary level of language teaching should focus on.

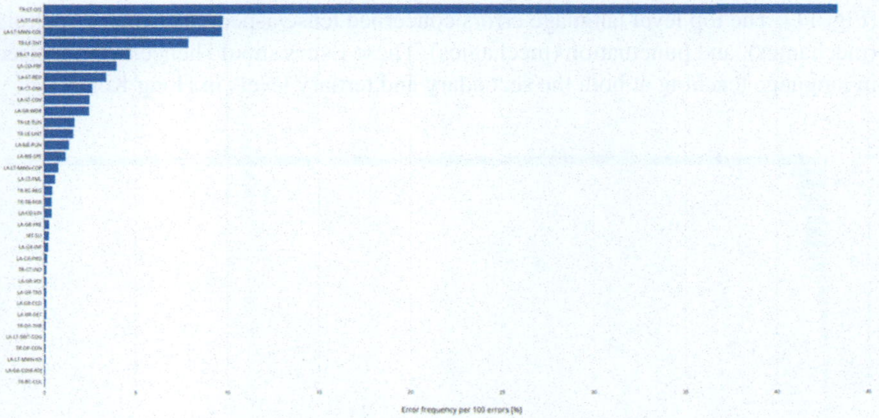

Fig. 8 Most frequent error tags in the Chinese sub-corpus

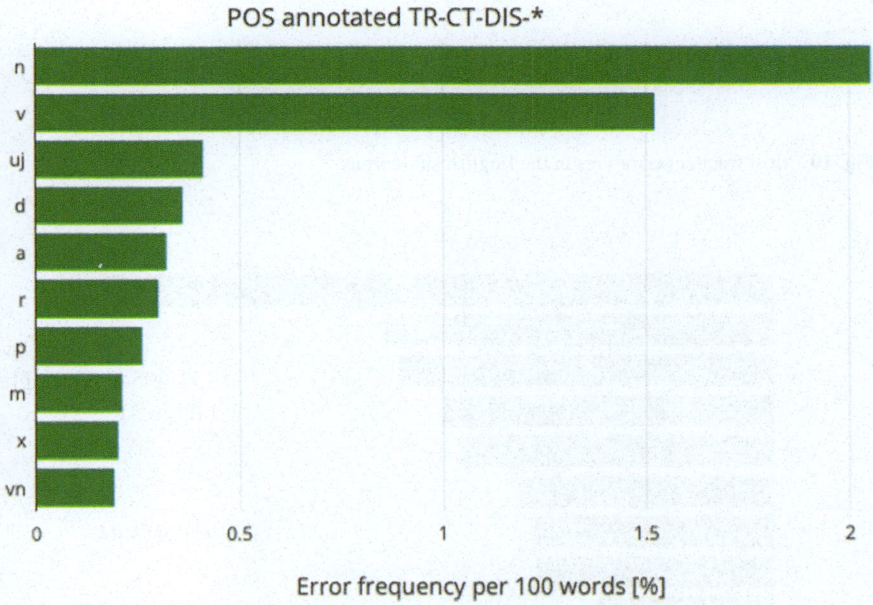

Fig. 9 POS annotated distortion in the Chinese sub-corpus

4.3 Most Frequent Error Tags in the English Sub-corpus

Figure 10 shows the most frequent error tags in the English sub-corpus. Likewise, distortion remained, apparently, the highest frequency error type. It was most frequently triggered by inaccurate target language expressions. Nouns and prepositions were amongst the most common parts-of-speech in which distortion occurred

(Fig. 11). The top level language errors concerned tense/aspect (grammar), spelling (mechanics), and punctuation (mechanics). These issues should hence be prioritised in language teaching at both the secondary and tertiary levels in Hong Kong.

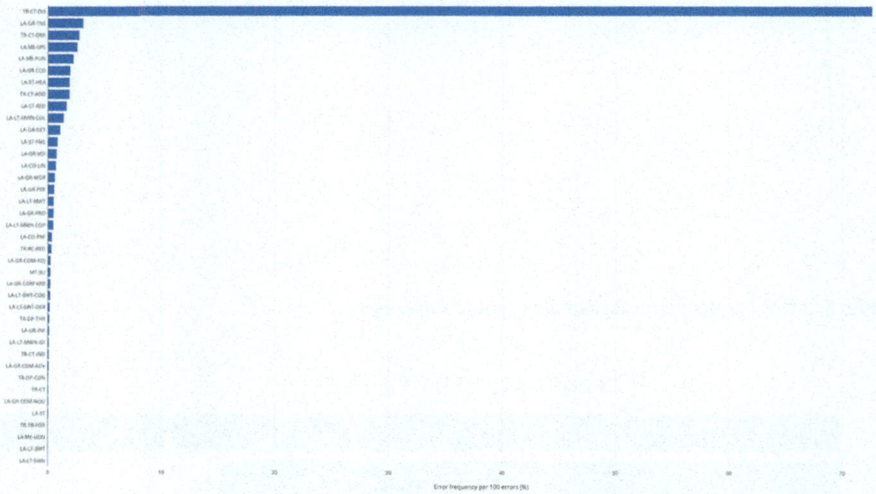

Fig. 10 Most frequent error tags in the English sub-corpus

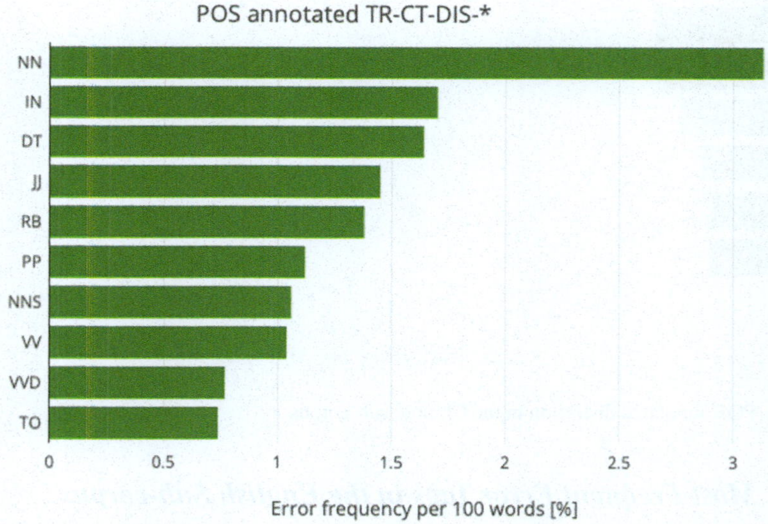

Fig. 11 POS annotated distortion in the English sub-corpus

	Word	Frequency ?		Word	Frequency ?
1	的 關 係	122	1	的 關 係	47
2	生物 多 樣性	107	2	關 係 較	26
3	生物 多樣 性	105	3	生物 多 樣性	25
4	野生 動物 的	103	4	封 電子 郵件	25
5	而 不 是	91	5	在 辦 公室	24
6	綠色 新 政	91	6	的 關 係 較	24
7	多 樣性 的	88	7	總統 選 舉	22
8	了 一 個	88	8	的 總統 選	22
9	生物 多 樣性 的	88	9	最 令 人	22
10	一 個 世紀	88	10	的 總統 選 舉	22
11	在 辦 公室	87	11	我們 的 情緒	22
12	我們 的 情緒	85	12	野生 動物 的	22

Fig. 12 Top 3- and 4-g of the student outputs (female [left] vs. male [right]) in the Chinese corpus

4.4 Gender and Students' Chinese/English Language Features

The typical 3- and 4-g of the students' outputs were computed using Sketch Engine. The results shown in Figs. 12 and 13 indicate that male and female students had slightly different preferences of word cluster use in both the Chinese and English outputs: female students tended to employ more regular phrases than male students did.

4.5 MOI and Students' Chinese/English Language Features

Likewise, the typical 3- and 4-g of the student outputs were compared between CMI and EMI students: EMI students seemed to employ slightly more regular phrases than CMI students did in general (Figs. 14 and 15).

	Word	Frequency ?	
1	Dream of Red	165	•••
2	of Red Mansions	161	•••
3	Dream of Red Mansions	153	•••
4	of the Red	135	•••
5	Dream of the	133	•••
6	the Red Chamber	133	•••
7	Dream of the Red	133	•••
8	of the Red Chamber	132	•••
9	told me to	130	•••
10	a blood test	105	•••
11	for two years	104	•••
12	for three months	103	•••

	Word	Frequency ?
1	Dream of Red	54
2	of Red Mansions	53
3	Dream of Red Mansions	53
4	Dream of the	37
5	Dream of the Red	37
6	of the Red	37
7	the Red Chamber	31
8	of the Red Chamber	31
9	told me to	30
10	stone is the	25
11	A Dream of	25
12	A Dream of Red	24

Fig. 13 Top 3- and 4-g of the student outputs (female [left] vs. male [right]) in the English corpus

	Word	Frequency ?
1	的 關 係	85
2	生物 多 樣性	67
3	而 不 是	63
4	生物 多樣 性	62
5	野生 動物 的	61
6	多 樣性 的	56
7	生物 多 樣性 的	56
8	封 電子 郵件	52
9	一 個 世紀	52
10	用 了 一	51
11	和 食物 鏈	51
12	我們 的 情緒	50

	Word	Frequency ?
1	的 關 係	84
2	生物 多樣 性	67
3	野生 動物 的	65
4	生物 多 樣性	65
5	在 辦 公室	62
6	總統 選 舉	60
7	了 一 個	59
8	一 個 世紀	58
9	的 總統 選	58
10	的 總統 選 舉	58
11	我們 的 情緒	57
12	綠色 新 政	56

Fig. 14 Top 3- and 4-g of the student outputs (CMI [left] vs. EMI [right] in secondary school) in the Chinese corpus

	Word	Frequency ?		Word	Frequency ?
1	Dream of Red	124	1	Dream of Red	95
2	of Red Mansions	120	2	of Red Mansions	94
3	Dream of Red Mansions	116	3	Dream of Red Mansions	90
4	of the Red	93	4	told me to	86
5	Dream of the Red	91	5	of the Red	83
6	Dream of the	91	6	Dream of the	83
7	the Red Chamber	87	7	Dream of the Red	83
8	of the Red Chamber	86	8	of the Red Chamber	81
9	told me to	75	9	the Red Chamber	81
10	for three months	69	10	for two years	62
11	for two years	67	11	a blood test	60
12	a blood test	65	12	my cholesterol level	58

Fig. 15 Top 3- and 4-g of the student outputs (CMI [left] vs. EMI [right] in secondary school) in the English corpus

4.6 *Previous Study Background and Chinese/English Language Features*

When students' previous study background (translation vs. non-translation) is taken into account, translation students seemed to use slightly more regular phrases than non-translation students did in general (Figs. 16 and 17).

4.7 *Language Proficiency and Chinese/English Language Features*

Finally, students' self-perceived target language proficiency (Native, Advanced vs. Intermediate) also seems to lead to different preferred word clusters: those of "native" target language performed slightly better than "advanced" and "intermediate" in producing more regular phrases (Figs. 18 and 19).

	Word	Frequency ?		Word	Frequency ?
1	最 佳 電影	12	1	的 關 係	165
2	的 製 作 人	11	2	生物 多 樣性	124
3	的 製 作	11	3	生物 多樣 性	121
4	製 作 人	11	4	野生 動物 的	117
5	美國 總 統	10	5	在 辦 公室	106
6	最 年輕 的	10	6	封 電子 郵件	105
7	的 最 佳	9	7	總統 選 舉	104
8	總 統 特朗普	9	8	而 不 是	103
9	野生 動物 的	9	9	我們 的 情緒	103
10	和 食物 鏈	8	10	一 個 世紀	103
11	海洋 和 食物	8	11	多 樣性 的	102
12	學生 辦 公室	8	12	的 總統 選 舉	101

Fig. 16 Top 3- and 4-g of the student outputs (translation [left] vs. non-translation [right]) in the Chinese corpus

5 Conclusions and Recommendations

This study aimed to investigate the Chinese and English language proficiency of tertiary students in Hong Kong through the unique lenses of translation. It identified the high-frequency error types in written Chinese/English of tertiary students in Hong Kong, and the relationship between Chinese/English language features and relevant contextual/learner factors.

The over 300,000-word error-annotated translation learner corpus developed in the study can provide rich research and teaching resources. Granger (1998) puts forward a significant factor of consideration in compiling a learner corpus:

> One factor which has a direct influence on the size of learner corpora is the degree of control exerted on the variables … and this in turn depends on the analyst's objectives … If the researcher is an SLA [Second Language Acquisition] specialist who wants to assess the part played by individual learner variables such as age, sex or task type, or if he[/she] wants to be in a position to carry out both cross-sectional and longitudinal studies, then he[/she] should give priority to the quality rather than the quantity of the data. (p. 11)

Since the corpus included learner, task and source text metadata of more than 30 types, and was annotated using a three-layer annotation scheme of over 40 error types, it is considered large in scale for the current study. As a matter of fact, the corpus constitutes the largest annotated subset within the entire MUST corpus at

	Word	Frequency		Word	Frequency
1	told me to	11	1	Dream of Red	217
2	what to do	9	2	of Red Mansions	212
3	for two years	9	3	Dream of Red Mansions	204
4	the time has	9	4	of the Red	173
5	If you can	8	5	Dream of the Red	171
6	me to take	8	6	Dream of the	171
7	my cholesterol level	8	7	the Red Chamber	165
8	the blood test	8	8	of the Red Chamber	164
9	one by one	8	9	told me to	150
10	for three months	8	10	for two years	120
11	I went to	7	11	a blood test	118
12	He told me to	7	12	for three months	115

Fig. 17 Top 3- and 4-g of the student outputs (translation [left] vs. non-translation [right]) in the English corpus

	Word	Frequency		Word	Frequency		Word	Frequency
1	的 關 係	167	1	額 外 的	28	1	額 外 的	17
2	生物 多 樣性	128	2	學生 辦 公室	23	2	貧 窮 學生	14
3	生物 多樣 性	123	3	外 的 表格	17	3	學生 辦 公室	12
4	野生 動物 的	122	4	小 時 的	17	4	外 的 壓力	11
5	在 辦 公室	111	5	額 外 的 表格	15	5	學生 填 寫 額	11
6	而 不 是	107	6	貧 困 學生	14	6	寫 額 外	11
7	我們 的 情緒	107	7	過 於 保守	14	7	學生 填 寫	11
8	封 電子 郵件	107	8	寫 額 外	14	8	小 時 的	11
9	一 個 世紀	106	9	填 寫 額	14	9	填 寫 額	11
10	多 樣性 的	106	10	填 寫 額 外	14	10	填 寫 額 外	11
11	總統 選 舉	105	11	聖 約翰 大學	14	11	的 學生 填	11
12	生物 多 樣性 的	105	12	大學 的 入學	13	12	大學 的 入學	11

Fig. 18 Top 3- and 4-g of the student outputs (native [left], advanced [middle] vs. intermediate [right]) in the Chinese corpus

	Word	Frequency		Word	Frequency		Word	Frequency
1	of the Red	117	1	of Red Mansions	113	1	Dream of Red	11
2	Dream of the Red	117	2	Dream of Red	113	2	the past ten	8
3	Dream of the	117	3	Dream of Red Mansions	113	3	of Red Mansions	7
4	of the Red Chamber	114	4	A Dream of	66	4	Dream of Red Mansions	7
5	the Red Chamber	114	5	A Dream of Red	65	5	the past ten years	7
6	Dream of Red	95	6	told me to	63	6	past ten years	7
7	told me to	94	7	for three months	60	7	but the time	6
8	of Red Mansions	94	8	of the Red	59	8	in the past	6
9	Dream of Red Mansions	86	9	for two years	58	9	time has passed	6
10	for two years	68	10	Dream of the	57	10	in the past ten	6
11	a blood test	67	11	Dream of the Red	57	11	spiritual stone is	5
12	for three months	62	12	the Red Chamber	54	12	one by one	5

Fig. 19 Top 3- and 4-g of the student outputs (native [left], advanced [middle] vs. intermediate [right]) in the English corpus

the time of writing this chapter. Moreover, with data collected from more than 11 institutions across Hong Kong, the corpus can be regarded representative of current language proficiencies of the target student population.

Based on results obtained from this large-size annotated corpus, the study has identified distortion as the highest frequency translational error of both Chinese–English and English–Chinese students in Hong Kong. This echoes the results obtained by Izquierdo et al. (2021) who, employing the same MUST TAS 1.0 annotation scheme on the translation of multiword expressions (22,184 words annotated), found a similar pattern of top rated error in their English to Spanish subset. At the time of writing this chapter, only a couple of annotated corpora have been uploaded by the MUST partners, with this Hong Kong subset being the largest annotated subset. In the future, it will be worthwhile to compare the results of the HKST dataset to annotated MUST subsets developed by partners in other language combinations, if available and of similar size. Collaborative interests have been expressed by a number of partners at the latest MUST Workshop, and opportunities for conducting comparative studies are currently being explored.

Within the categories of language errors, the top three problems in students' English–Chinese translation were heavy structure (style and situational context), multiword non-term collocation (lexis and terminology), and pronoun reference (cohesion), and the top three in Chinese–English concerned tense/aspect (grammar), spelling (mechanics), and punctuation (mechanics). These issues have been identified as the most pressing problems that require attention in both secondary and tertiary language teaching levels.

The error batteries used to annotate the Chinese and English problems of tertiary students in the HKST corpus of this study consisted of both translational and linguistic errors. By using translation as a lens to study students' language proficiency, the study supports the idea that translation tasks can provide valuable insights into students' understanding of both the target language and the source language. This approach can reveal the ways in which students' comprehension and use of the source language may influence their performance in the target language. Thus, the use of error batteries that include translational errors can help to create a more comprehensive and nuanced understanding of students' language abilities and challenges.

The error batteries can also be utilised for teaching and assessment purposes. The annotated learner corpus developed in this study provides examples of different error types that can be used to train students. Furthermore, language and translation teachers can use the annotation scheme in their assessments, providing students with feedback based on the frequency of annotations made on their translations. By using the error batteries in this way, students can gain a better understanding of the specific errors they make and how to address them, ultimately improving their language proficiency.

The study also suggests that learner factors, such as gender, MOI at secondary schools, previous study background, and self-perceived language proficiency, may result in variations in the language features observed in both Chinese and English translational outputs produced by students. To address these findings, we recommend the development of tailor-made exercises to enhance identified deficiencies in students' Chinese and English writing. In addition, dedicated online platforms can be created to showcase students' errors in translational Chinese/English writing and provide pedagogical solutions to these errors (cf. Pan et al. 2022). By addressing students' specific needs in this way, we can help to improve their language proficiency and better prepare them for academic and professional success.

To conclude, the study, with its rich annotated data and student/context information collected, can provide valuable insight into the language proficiency, and most importantly, deficiencies of students in Hong Kong. Based on the findings obtained, more in-depth analyses can be carried out to find out the specific differences amongst learners of different language needs, and hopefully, longitudinal variances amongst learners based on pedagogical interventions. The corpus of this study will also be extended with the inclusion of more data. In addition, the study can be expanded in the near future to cover comparisons with existing and future learner corpora of other language combinations, especially those developed by other regional MUST partners who employ the same annotations scheme.

Acknowledgements The study is supported by the Language Fund under Research and Development Projects 2018–19 of the Standing Committee on Language Education and Research (SCOLAR), Hong Kong SAR. An earlier draft of this chapter was submitted to the funding body as part of the final project report.

References

Al Khafaji, H.A. Adil. 2007. Translanguage. *Meta* 52 (3): 436–476.

Behrens, Heike. 2008. *Corpora in Language Acquisition Research: History, Methods, Perspectives*. Amsterdam: John Benjamins.

Canale, Michael, and Merrill Swain. 1980. Theoretical bases of communicative approaches to second language teaching and testing. *Applied Linguistics* 1: 1–47.

Chan, Shelby Kar-yan, and Gilbert C. F. Fong. 2016. Hong Kong speak: Cantonese and Rupert Chan's translated theatre. In *The Oxford Handbook of Modern Chinese Literatures*, eds. Rojas Carlos, and Bachner Andrea. London: Oxford University Press.

Chomsky, Noam. 1965. *Aspects of the Theory of Syntax*. Cambridge: MIT Press.

Chow, Ian C., and Billy Tak-Ming Wong. 2015. The mega-sized, multi-genre Chinese-English parallel corpus for computer-aided translation. In *The International Conference on New Horizons in Translation Technology*. Hong Kong, China, 24 April.

Cook, Guy. 2010. *Translation in Language Teaching: An Argument for Reassessment*. Oxford: Oxford University Press.

Corder, Stephen Pit. 1967/1983. The significance of learners' errors. In *Second Language Learning: Contrastive Analysis, Error Analysis, and Related Aspects*, eds. Robinett Betty Wallace, and Schachter Jacquelyn, 163–172. Ann Arbor: The University of Michigan Press.

Evans, Stephen. 2013. The long march to biliteracy and trilingualism: Language policy in Hong Kong education since the handover. *Annual Review of Applied Linguistics* 33: 302–324.

Education Bureau. 2010. Fine-tuning the medium of instruction in high schools. http://www.edb.gov.hk/attachment/en/edu-system/primary-secondary/applicable-tosecondary/moi/2nd_moi_booklet.pdf

Fan, May Y. 2001. An Investigation into the vocabulary needs of university students in Hong Kong. *Asian Journal of English Language Teaching* 11: 69–85.

Fictumová, Jarmila, Adam Obrusnik, and Krystina Štěpánková. 2017. Teaching specialised translation. Error-tagged translation learner Corpora. *Sendebar* 28: 209–241.

Granger, Sylviane. 1998. *Learner English on Computer*. London: Longman.

Granger, Sylviane. 2002. A bird's eye view of learner corpus research. In *Computer Learner Corpora, Second Language Acquisition and Foreign Language Teaching*, ed. Granger Sylviane, Hung Joseph, and Petch-Tyson. Stephanie, 3–33. Amsterdam: John Benjamins.

Granger, Sylviane. 2003. The international corpus of learner English: A new resource for foreign language learning and teaching and second language acquisition research. *TESOL Quarterly* 37 (3): 538–546.

Granger, Sylviane, and Marie-Aude Lefer. 2017. *General Report of the MUST Kickoff Meeting*. Louvain-la-Neuve: Centre for English Corpus Linguistics, Université catholique de Louvain.

Granger, Sylviane, and Marie-Aude Lefer. 2020. The multilingual student translation corpus: A resource for translation teaching and research. *Language Resources and Evaluation* 54: 1183–1199.

Granger, Sylviane, Estelle Dagneaux, Fanny Meunier, and Magali Paquot, eds. 2009. *International Corpus of Learner English: Version 2*. Louvain-La-Neuve: Presses Universitaires de Louvain.

Gui, Shichun, and Huizhong Yang, eds. 2003. *CLEC—Chinese Learner English Corpus*. Shanghai: Shanghai Foreign Language Education Press.

Horner, Bruce, and Min-Zhan Lu. 2012. (Re) Writing English: Putting English in translation. In *English: A Changing Medium for Education*, eds. Leung Constant, and Brian V. Street. Bristol: Multilingual Matters.

Hymes, Dell. 1972. On communicative competence. In *Sociolinguistics: Selected Readings*, eds. J. B. Pride, and Holmes Janet, 269–293. Harmondsworth: Penguin.

Izquierdo, Marlén, Zurine Sanz, Naroa Zubillaga, and Elizabete Manterola. 2021. *Basque in Student Translations: What MUST Tell Us*. Paper presented at MUST Workshop: Université catholique de Louvain, Beligum (virtual conference).

Kilgarriff, Adam, Vít. Baisa, Jan Bušta, Miloš Jakubíček, Vojtěch Kovář, Jan Michelfeit, Pavel Rychlý, and Vít. Suchomel. 2014. The sketch engine: Ten years on. *Lexicography* 1: 7–36.
Laviosa, Sara. 2014. *Translation and Language Education: Pedagogic Approaches Explored*. London: Routledge.
Lin, Angel M. Y. 2015. Conceptualising the potential role of L1 in CLIL. *Language, Culture and Curriculum* 28 (1): 74–89.
Lin, Angel M. Y., and Evelyn Y. F. Man. 2009. *Bilingual Education: Southeast Asian Perspectives*. Hong Kong: Hong Kong University Press.
Lin, Linda H. F., and Bruce Morrison. 2010. The impact of the medium of instruction in Hong Kong secondary schools on tertiary students' vocabulary. *Journal of English for Academic Purposes* 9 (4): 255–266.
Lo, Yuen Y., and Eric Siu Chung Lo. 2014. A meta-analysis of the effectiveness of English-medium education in Hong Kong. *Review of Educational Research* 84 (1): 47–73.
Malmkjær, Kirsten. 1998. Introduction: Translation and language teaching. In *Translation and Language Teaching: Language Teaching and Translation*, ed. Malmkjær Kirsten, 1–11. Manchester: St. Jerome.
McEnery, Tony, Richard Xiao, and Yukio Tono. 2006. *Corpus-Based Language Studies: An Advanced Resource Book*. New York: Routledge.
Naimushin Boris. 2002. Translation in foreign language teaching: The fifth skill. *Modern English Teacher* 11 (4): 46–49.
Ngan, Heltan Y. W. 2009. Developing biliteracy through studying the bilingual representation phenomenon in translation texts. *Babel* 55 (1): 40–57.
Obrusník, Adam 2014. Hypal: A user-friendly tool for automatic parallel text alignment and error tagging. In *11th International Conference Teaching and Language Corpora*, Lancaster, 20–23 July 2014, 67–69.
PACTE. 2003. Building a translation competence model. In *Triangulating Translation: Perspectives in Process Oriented Research*, ed. Alves Fábio, 43–66. Amsterdam: John Benjamins.
Pan, Jun. 2012. *Problem analysis and the learning of interpreting: Perceptions, evaluation and corpus analysis of students' interpreting work (Ph.D. dissertation)*. City University of Hong Kong, Hong Kong.
Pan, Jun. 2014. Repetition and self-correction in students' interpreting performance: Corpus evidence of the "why" and "how". In *The 4th Using Corpora in Contrastive and Translation Studies Conference*, Lancaster, United Kingdom, 24–26 Jul.
Pan, Jun. 2017. A corpus-based study of college students' translation performance: The construction and initial findings of the HK-CL(CE/EC)TC. In *General Report of the MUST Kickoff Meeting*, ed. Sylviane Granger, and Lefer Marie-Aude, 147–168. Louvain-la-Neuve: Centre for English Corpus Linguistics, Université catholique de Louvain.
Pan, Jun. 2019a. Researching translator and interpreter training: Convergences and divergences. Invited keynote speech presented at the Guangdong-Hong Kong-Macau Postgraduate Academic Exchanges in Foreign Languages and Translation, Sun Yat-Sen University, Zhuhai, 11–13 May.
Pan, Jun. 2019b. Employing learner corpora in the study of translator and interpreter training: Implications from lexical cohesion. Paper presented at The Ewha GSTI Conference 2019: Science and Technology in Translation and Interpreting, Seoul, Korean, 9 November. In *Ewha GSTI Conference Proceedings*, 39.
Pan, Jun. 2021a. Researching translator and interpreter training: convergences and divergences. Invited talk by School of Foreign Language Studies, Zhejiang Scie-Tech University (virtual seminar, 31 Mar).
Pan, Jun. 2021b. Translator and (or versus?) interpreter training—Topics, methods, and empirical findings Researching translator and interpreter training: Convergences and divergences. Invited talk by Division of Humanities and Social Sciences of BNU-HKBU United International College (virtual seminar, 6 May).

Pan, Jun, and Jackie Xiu Yan. 2012. Learner variables and problems perceived by students: An investigation of a college interpreting programme in China. *Perspectives: Studies in Translatology* 20(2):199–218.

Pan, Jun, and Honghua Wang. 2012. Investigating the nature of the semi-natural interpretation: A case study. In *Interpreting Brian Harris: Recent Developments in Translatology*, eds. María Amparo, Jiménez Ivars, and María Jesús, Blasco Mayor, 77–94. Switzerland: Perter Lang.

Pan, Jun, and Shelby Kar-yan Chan. 2013. Investigating the routes to professional translators/interpreters: The construction and development of the HK-CL(CE/EC)TIC. In *The 2nd Business Translation Forum of China*, Beijing, PRC, 25–26 May.

Pan, Jun, and Jackie Xiu Yan. 2014. Inaccurate pronunciation in students' interpreting performance: Evidence from a learner corpus. In *The 11th Teaching and Language Corpora Conference*, Lancaster, United Kingdom, 20–23 Jul.

Pan, Jun, and Billy Tak-Ming Wong. 2015a. Pragmatic markers in interpreted political discourse: A corpus-driven study. In *The International Conference on Corpus Linguistics and Technology Advancement (CoLTA)*, Hong Kong, 16–18 Dec.

Pan, Jun, and Billy Tak-Ming Wong. 2015b. Investigating pragmatic markers in interpreted political speeches from Chinese to English. In *The International Conference "Found in Translation—Translations are the Children of their Times"*, Bucharest, Romania, 10–11 Sep.

Pan, Jun, and Honghua Wang. 2017. The development of textual competence in student translators: A corpus-based study of problems of coherence and cohesion. In *Translation in Transition 3 (TT3)*, Ghent, Belgium, 13–14 Jul.

Pan, Jun, and Billy Tak-Ming Wong. 2017. Developing pragmatic competence in political retour interpreting: A corpus-driven study on the use of pragmatic markers. In *The Teaching Translation and Interpreting Conference*, Łódź, Poland, 15–16 Sep.

Pan, Jun, and Honghua Wang. 2018. Learner factors relating to errors of coherence and cohesion in translation: Some preliminary findings. Paper presented at the MUST (Multilingual Student Translation) Workshop, Université catholique de Louvain, Belgium, 11 Sep.

Pan, Jun, and Billy Tak-Ming Wong. 2021. Distortion in student translations: Annotation of the Hong Kong subset of the MUST corpus. Paper presented at MUST Workshop. Université catholique de Louvain, Beligum (virtual conference), 18 Nov.

Pan, Jun, Billy Tak-Ming Wong, Shelby Kar-yan Chan, and Honghua Wang. 2021a. Investigating the Chinese and English language proficiency of tertiary students in Hong Kong: Perspectives from the Hong Kong Subset of the multilingual student translation corpus. Invited paper at the 25th Anniversary Conference of the Standing Committee on Language Education and Research (SCOLAR). In *Programme Book*, 4. The Hong Kong Convention and Exhibition Centre, Hong Kong, 25 June.

Pan, Jun, Billy Tak-Ming Wong, and Honghua Wang. 2021b. Making a way through the jungle: Exploring learner data in translator and interpreter training. Plenary paper at the International Symposium on Corpora and Translation Education. In *Programme Book*, 15–17. Hong Kong Baptist University, Hong Kong (virtual conference). 5–6 June.

Pan, Jun, Billy Tak-Ming Wong, and Honghua Wang. 2022. Navigating learner data in translator and interpreter training. *Babel* 68 (2): 236–266.

Poulisse, Nanda. 1999. *Slips of the Tongue: Speech Errors in First and Second Language Production.* Amsterdam/Philadelphia: John Benjamins.

Ricardo-Osorio, José G. 2008. A study of foreign language learning outcomes assessment in U.S. undergraduate education. *Foreign Language Annals* 41(4): 590–610.

Richards, Jack C., and Theodore S. Rodgers. 2001. *Approaches and Methods in Language Teaching.* Cambridge: Cambridge University Press.

Selinker, Larry. 1972/1983. Interlanguage. In *Second Language Learning: Contrastive Analysis, Error Analysis, and Related Aspects*, eds. Robinett Betty Wallace, and Schachter Jacquelyn, 173–196. Ann Arbor: The University of Michigan Press.

Selinker, Larry. 1992. *Rediscovering Interlanguage.* London: Longman.

Shih, Hsue-Hueh. 2000. Compiling Taiwanese learner corpus of English. *Computational Linguistics and Chinese Language Processing* 5 (2): 87–100.

Sidiropoulou, Maria. 2015. Translanguaging aspects of modality: Teaching perspectives through parallel data. *Translation and Translanguaging in Multilingual Contexts* 1 (1): 27–48.

Steward, Dominic, Silvia Bernardini, and Aston Guy. 2004. Introduction: Ten years of TaLC. In *Corpora and Language Learners*, ed. Aston Guy, Bernardini Silvia, and Steward Dominic, 1–20. Amsterdam: John Benjamins.

Sun, Youyi, and Liying Cheng. 2013. Assessing second/foreign language competence using translation: The case of the college English test in China. In *Translation in Language Teaching and Assessment*, ed. Tsagari Dina and Floros Georgios, 235–252. Newcastle upon Tyne: Cambridge Scholars Publishing.

Tsagari, Dina, and Georgios Floros, eds. 2013. *Translation in Language Teaching and Assessment*. Newcastle upon Tyne: Cambridge Scholars Publishing.

Tsang, Wing-Kwong. 2008. Evaluation research on the implementation of the medium of instruction guidance for secondary schools. *HKIED Research Newsletter* 24: 1–7.

Wen, Qiufang, Lifei Wang, and Maocheng Liang, eds. 2005. *SWECCL—Spoken and Written English Corpus of Chinese Learners*. Beijing: Foreign Language Teaching and Research Press.

Wong, Billy Tak-Ming. 2010. Semantic evaluation of machine translation. In *The 7th International Conference on Language Resource and Evaluation (LREC)*, 2884–2888. Valletta, Malta, 19–21 May.

Wong, Billy Tak-Ming, and Sophia Y. M. Lee. 2013. Annotating legitimate disagreement in corpus construction. In *The 11th Workshop on Asian Language Resources (ALR)*, 51–57. Nagoya, Japan, 14 Oct.

Wong, Billy Tak-Ming, Ian C. Chow, Jonathan Webster, and Hengbin Yan. 2014. The Halliday Centre Tagger: An online platform for semi-automatic text annotation and analysis. In *The 9th International Conference on Language Resources and Evaluation (LREC)*, 1664–1667. Reykjavik, Iceland, 26–31 May.

Yan, Xiu Jackie, and Honghua Wang. 2012. Second language writing anxiety and translation: Performance in a Hong Kong tertiary translation class. *The Interpreter and Translator Trainer* 6 (2): 171–194.

Yan, Xiu Jackie, and Honghua Wang. 2014. The construction and application of an error annotated learner translation corpus in translation classes. In *11th International Conference Teaching and Language Corpora*, Lancaster, UK, 20–23 Jul.

Yan, Xiu Jackie, and Honghua Wang. 2015. The interplay between software usage, motivation and gender differences: A survey based on a Putonghua classroom in Hong Kong. *Overseas Chinese Education* 76 (3): 368–376.

Yan, Xiu Jackie, and Jun Pan. 2016. Backgrounds, attitudes and software application of tertiary-level Putonghua learners in Hong Kong: A focus group interview study (in Chinese). *Journal of International Chinese Studies* 7 (1): 176–188.

Yan, Xiu Jackie, Jun Pan, and Honghua Wang. 2010. Learner factors, self-perceived language ability and interpreting learning: An investigation of Hong Kong tertiary interpreting classes. *The Interpreter and Translator Trainer* 4 (2): 173–196.

Yang, Huizhong, and Naixing Wei, eds. 2005. *COLSEC—College Learners' Spoken English Corpus*. Shanghai: Shanghai Foreign Language Education Press.

Jun Pan is Associate Professor in the Department of Translation, Interpreting, and Intercultural Studies at Hong Kong Baptist University, where she also holds the positions of Associate Dean (Research) of the Faculty of Arts and Associate Head of the Department. She serves as Co-editor of *Bandung: Journal of the Global South* and Review Editor of *The Interpreter and Translator Trainer*. Her research interests lie in learner factors in interpreter training, corpus-based interpreting/translation studies, digital humanities and interpreting/translation, interpreting/translation and political discourse, professionalism in interpreting, etc. Her recent work includes a

6.5-million-word corpus on Chinese/English political interpreting and translation (https://digital.lib.hkbu.edu.hk/cepic/). Dr. Pan is also President of the Hong Kong Translation Society.

Billy Tak Ming Wong is Senior Research Coordinator of Hong Kong Metropolitan University. His research interests lie in language technology and the use of technology in education. He has been teaching and conducting research on computer-aided translation and technology-enhanced education for more than 15 years, and has widely published on evaluation of machine translation quality and the impacts of technology on education.

Honghua Wang is Assistant Professor in the School of Translation and Foreign Languages at The Hang Seng University of Hong Kong. Her research interests include interpreter and translator training, gender and translation, and second language acquisition. She has many years of teaching experience and published widely in internationally renowned journals.

The manufacturer's authorised representative in the EU is Springer Nature Customer Service Centre GmbH, Europaplatz 3, 69115 Heidelberg, Germany. If you have any concerns regarding our products, please contact ProductSafety@springernature.com

Printed and bound by CPI Group (UK) Ltd, Croydon, CR0 4YY
01/10/2026
02225877-0018